NEW IDEAS IN ACCELERATOR TECHNOLOGY

Electromagnetic accelerators of the future

S. N. DOLYA

Dubna, 2014

The book presents articles by a well-known accelerator-physicist S. N. Dolya of the Joint Institute for Nuclear Research, Dubna, Russia. The articles are devoted to acceleration of physical bodies to hypersonic (5 km/s) and space (8.5 km/s) velocities. The passage by these bodies through the Earth's atmosphere has been calculated. It is demonstrated that the flight range of such accelerated bodies can reach several thousand kilometers, the lift above the Earth's surface - thousands of kilometers.

It is possible to use such accelerators for diverse purposes. If you install such an accelerator on a satellite, the accelerated bodies, micrometeorites, would reach the Earth's surface without disintegration.

If you install a radio transmitter on a body, shaped as a bicycle spoke, and send this body to the cometary nucleus area, you can determine, based on signals from these bodies, the size of the comet's nucleus. Through irradiation of a deuterium - tritium mixture with such bodies it is possible to create an accelerator-driven fusion reactor.

The annex discusses the possibility of a gas-dynamic acceleration of bodies to hypersonic speeds. The flight range of such a body with a mass of 1 kg reaches 1400 km, flight altitude -350 km.

An attentive reader will find a lot of interesting things in this book.

Content

I. Acceleration of electrically charged bodies

II. Acceleration of the magnetic and electric dipoles

III. Application accelerators

Appendix

I. Acceleration of electrically charged bodies

About the electrodynamics acceleration of cylinder-shaped particles

A possibility of electrodynamic acceleration of particles from the initial zero velocity to the final velocity $V_{fin} = 10$ km/s over the acceleration length $L_{acc} = 5$ m is considered. After the electrostatic preacceleration to $V_{in} = 1$ km/s particles are accelerated at the trailing edge of the voltage pulse $U_{pulse} = 5.75$ MV, which runs along the spiral turns. Accelerated particles have the mass $m = 6 * 10^{-6}$ g, diameter $d_{sh} = 6$ μ, and length $l_{sh} = 1$ cm. Because of a pointed cone at the head the particles can move in the air almost without loss of velocity, penetrating into aluminum and water as deep as $l_{Al} = 10$ cm and $l_{water} = 1$ m respectively.

Introduction

It is known [1] that magnetic dipoles can be accelerated by orienting them in space so that the magnetization axis coincides with the acceleration axis and applying the current pulse field.

Being a difference interaction, the dipole interaction is inherently much less effective than the monopole charge interaction. For example, if one of the poles of the dipole is repelled by the accelerating pulse, the other is at the time is attracted to the current pulse and the resulting accelerating force is the difference of these two forces. Therefore, a uniform magnetic field does not act on the dipoles; the dipoles can be accelerated by the magnetic field gradient.

In addition, iron, which is the most suitable material for acceleration, has a low specific magnetic moment (magnetic moment per nucleon):
$m = 2 * 10^{-10}$ eV / (Gs * nucleon). Accordingly, the energy gain rate of the iron magnetic dipole: $m * \partial B / \partial z$ in a gradient magnetic field with a reasonable gradient $\partial B / \partial z = 500$ Gs / cm is found to be
$\Delta W_{mag} = m * \partial B / \partial z = 2 * 10^{-10} * 500 = 10^{-7}$ eV / cm.

Energy gain by a particle having a specific charge (the charge per nucleon in units of electron charge) can amount to e $(Z / A) = 10^{-8}$, and in the electric field $E = 10$ kV / cm the energy gain rate is found to be
$\Delta W_{el} = 10^{-8} * 10^4 = 10^{-4}$ eV / cm, which is three orders of magnitude higher than the energy gain rate in the dipole interaction.

The spherical shape of the particles used in some systems is not optimal for many reasons, especially acceleration-related ones. As the radius of particle increases, the acceleration efficiency dramatically drops. To understand better the fundamental shortcomings of acceleration of spherical particles, let us compare the main parameters of accelerated iron balls with different diameter.

Table 1. The main parameters of the accelerated objects

D, μ	A	Z	Z/A	eΦ, MeV	M, g	β_i
2	$2*10^{13}$	$6*10^7$	$3*10^{-6}$	0.1	$3.2*10^{-11}$	$4*10^{-5}$
20	$2*10^{16}$	$6*10^9$	$3*10^{-7}$	1	$3.2*10^{-8}$	$1.2*10^{-5}$
200	$2*10^{19}$	$6*10^{11}$	$3*10^{-8}$	10	$3.2*10^{-5}$	$4*10^{-6}$
$2*10^3$	$2*10^{22}$	$6*10^{13}$	$3*10^{-9}$	10^2	$3.2*10^{-2}$	$1.2*10^{-6}$

In all cases the electric field on the surface of the object is $E_{surf} = 10^9$ V / cm. The first column is the diameter of the ball in μm, the second column is the atomic weight of the ball in units of the atomic mass of the nucleon, the third column is the limiting charge Z on the object in units of the electron charge, the fourth column is the ratio of the charge on the ball to its weight, the fifth column is the potential Φ of the object (the energy that an electron must have to overcome the repulsion of the electrons previously placed on the object), the sixth column is the mass of the ball in grams, and the seventh column is the initial velocity of the balls acquired by them after the acceleration in the electrostatic field with the voltage $U_{inj} = 250$ kV, expressed in units of the speed of light $\beta_{in} = V / c$, where $c = 3 * 10^5$ km / s is the velocity of light in vacuum.

A comparison of the data presented in Table 1 shows that as the diameter of the balls increases, their weight and atomic mass (column 6) increase as the cube of the radius, and the charge to be placed on the ball to get the field strength $E_{surf} = 10^9$ V / cm increases as the square of the radius. For a ball with the diameter D = 2 mm the charge (in units of electron charge) $Q = I * \tau = 2$ A * 5 μs = $6 * 10^{13}$ is already close to the limiting charge per pulse accelerated in linear accelerators. The ratio of the charge placed on the ball to the mass of the ball (column 4) linearly decreases with increasing diameter, which means that the acceleration efficiency linearly with increasing diameter, i.e., balls of a larger diameter will gain less velocity over the same accelerator length and at the same intensity.

For cylinder-shaped particle the parameter Z / A does not depend on the length of the cylinder at all because the charge placed on the particle and its mass linearly increase with increasing length of the cylinder. Platinum coating and oxygen passivation of the particle greatly increases its surface barrier preventing electrons from leaking from with particles.

The cylindrical shape of particles with a pointed cone at the head allows much better ballistic performance than that of the prototype, namely, the pointed cone ensures a very low drag coefficient, which is important for moving through the atmosphere without losing velocity. The use of tungsten instead of iron significantly improves the aerodynamic "quality" of particles.

Finally, elongated cylindrical particles will be able to penetrate into a material many times deeper than spherical particles, which is of interest for some practical applications.

Accelerator

1. Introduction of 3.3% of iron in tungsten

Let us add 3.3% of iron to a tungsten particle and magnetize this macroparticle so that the axis of magnetization coincides with the longitudinal axis of the cylinder and the axis of acceleration. Since the atomic weight of tungsten is about 3.3 times greater than that of iron and tungsten does not has intrinsic magnetic moment, the resulting specific magnetic moment will be approximately 100 times lower than that of iron:
$m_1 = 2 * 10^{-12}$ eV / (Gs * nucleon).

2. Platinum coating and oxygen passivation of the tungsten wire

To create a surface barrier for electrons "accommodated" on the particle, the work function of the electrons must be increased as much as possible. Best-known is the work function of platinum passivated with oxygen [2].
The charge located on the macroparticle will leak from it through field emission according to the formula [2],

$$j = e^2E^2 / (8\pi h\varphi) * \exp \{[- (8\pi / 3) (2m)^{1/2}/h] * [(e\varphi)^{3/2} / (eE) * \theta (y)]\}, (1)$$

where $\theta (y)$ is the Nordheim function in which the argument is a relative reduction in the work function of an external electric field in the sense of Schottky.

3. Spatial orientation and pre-acceleration of particles

We estimate the velocity of particles accelerated by a coil in which the current is turned off when the macroparticle reaches the middle of the coil. The field strength on the axis of the coil is described by the formula

$$B_z = [2\pi I_\varphi / c] * R_0^2 / [z^2 + R_0^2]^{3/2}, \qquad (2)$$

where R_0 is the radius of the current coil turn, I_φ is the current in the turn, and z is the distance from the center of the turn.

Let the particle be pre-magnetized and have a specific magnetic moment two orders of magnitude lower than that of iron, namely, $m_1 = 2 * 10^{-12}$ eV / (Gs * nucleon). Using formula (2), we find the magnetic field gradient $\partial B_z / \partial z = B_z * 3z / [z^2 + R_0^2]$ and the particle energy increment resulting from the interaction of the particle magnetic moment with the magnetic field gradient of the current turn

$$\Delta W_{in1} = m_1 * (\partial B_z / \partial z) * l_{acc}, \qquad (3)$$

which for $l_{acc} = 10$ cm and the magnetic field gradient $(\partial B_z / \partial z) = 500$ Gs / cm can be estimated at $\Delta W_{in1} = 10^{-8}$ eV / nucleon, so that the velocity of the particles after pre-acceleration in this turn (coil gun) will be $V_{in1} = 1.5$ m / s and the segment will be directed along the axis of acceleration.

4. Electron beam irradiation of particles: Electron energy

Thus, we have slightly accelerated the macroparticle due to its weak magnetization and, which is important, oriented it along the axis of acceleration. Further acceleration of the particle will be due to its interaction with the electric field; therefore, it must be an electrically charged macroparticle. The "charge" can be produced by irradiating the macroparticle with an electron beam.

The surface strength of the field is takes to be $E_{surf} = 40$ MV / cm (later we will analyze in detail the choice of this very important parameter). Then, for the wire diameter $d = 6$ μ we find that the minimum energy of the electrons capable of overcoming the Coulomb repulsion of the electrons previously placed on the macroparticle should be $W_e > eE_{surf} * d_{sh} / 2 = 12$ keV.

5. Electron beam irradiation of particles: Mean free path

The path of electrons with an energy of 12 keV in aluminum is approximately 2 mg/cm² [2]. Given the density of aluminum ρ_{Al} = 2.7 g/cm³, we find that the extrapolated path of electrons in aluminum is $l_{Al} \approx 7$ µ. Since the tungsten density is ρ_{tung} = 19.35 g/cm³, more than 7 times higher than the density of aluminum, the mean free path of electrons with an energy of 12 keV in tungsten is approximately 1 µ. Thus, electrons with the energy W_e = 12 keV are not able to cross the particle diameter 6 µ, will lose their energy by ionization of matter, and will stop deep in the particle.

6. Electron beam irradiation of particles: Irradiating beam density

We estimate the time of flight of the macroparticle over a distance equal to its own length, that is, l_{sh} = 1 cm. At the particle velocity V_{sh} = 1.5 m / s this time will be $t_{sh} = l_{sh} / V_{sh}$ = 7 ms. Let the duration of the electron beam irradiation be t_{irr} = 300 µs, Below we will show that during this time the macroparticle fully charged electrons will lose less than one percent of electrons through field emission. Let the total number of electrons placed on the particle be $N_e = 4 * 10^{10}$.

Let the transverse dimensions of the irradiating beam ribbon be $S_{irr} = 5\mu * 1cm$, where the beam cross section 5µ is chosen such that the length of the chord drawn through the cross section of the particle is greater than 1 µ, the depth of the extrapolated path of the electrons with the energy W_e = 12 keV in tungsten. Considering that the reflectance of slow electrons from tungsten is about k_{irr} = 0.35 [2], we find from the relation

$$j_{irr} * 6 * 10^{18} * S_{irr} * t_{irr} * k_{irr} = N_e \qquad (4)$$

that the desired current density of the electron beam irradiation is approximately j_{irr} = 0.1 A/cm², which is much lower than, for example, the electron emission current density obtained from oxide cathodes.

7. Electron beam irradiation of particles: Introduced pulse

We estimate the transverse particle velocity arising from the pulse introduced by electrons. The velocity of the electrons irradiating the particle is $\beta_e = V_e / c = 5 * 10^{-3}$, where $c = 3 * 10^{10}$ is the speed of light in vacuum. The transverse velocity of the particles resulting from electron irradiation will be

$$V_\perp = (N_e / A_a N_a) * m_e V_e / M_n, \qquad (5)$$

where N_e is the number of uncompensated (planted on macroparticle) electrons, N_a is the number of atoms in the particle, A_a is the atomic weight of tungsten, m_e is the electron mass, and M_n is the mass of the nucleon. If there are few excess electrons in the macroparticle $(N_e / A_a N_a) = (Z / A) = 1.23 * 10^{-8}$ and the electron-to-nucleon mass ratio is 1/2000, the transverse velocity is low. If it is a problem, particles will have to be irradiated with the two colliding electron beams.

8. Electron beam irradiation of particles: Field electron emission

To plant several charges on a macroparticle is not a problem, but when there are many electrons on the macroparticle, they will begin to leak from it through field emission. Let the field strength for field emission be $E = 4 * 10^7$ V / cm. After this field strength is attained, all the newly planted electrons will leak from the particles due to the Coulomb repulsion.

Once there is quite a lot of electrons planted on the particle, it becomes necessary to overcome their repulsion for planting more electrons. This means that the energy of electrons that we want to plant on the particle should be high enough to allow them to overcome this Coulomb barrier, reach the particle, and stay on it.

For spherical particles the Coulomb barrier linearly increases with increasing radius, and for particles with the diameter $d = 2$ mm it can be as high as tens of MeV, which will require a special accelerator for the electrons to overcome it.

Field emission will be an obstacle for planting a high electric charge on the particle. Part of the charge will continuously leak from particles due to the tunnel effect. The dependence of the field-emission current density on the electric field strength for surfaces with different work functions is given in [2].

We take the following parameters: density $\rho_{tung} = 19$ g/cm^3, particle diameter $d_{sh} = 6$ μ, length $l_{sh} = 1$ cm, particle cross-section $S_{tr} = \pi d^2 / 4 = 2.8 * 10^{-7}$ cm^2, volume $V_{sh} = 2.8 * 10^{-7}$ cm^3, the mass of the tungsten wire segment (particle) $m_{sh} = 6 * 10^{-6}$ g, lateral particle surface $S_{surf} = \pi d_{sh} * l_{sh} = 10^{-3}$ cm^2, and surface tension of the field on the macroparticle $E_{surf} = 4 * 10^7$ V / cm.

Let us find the charge on the particle that is enough to create the electric field $E_{surf} = 4 * 10^7$ V / cm from the expression relating the strength of the field on a cylinder to the linear charge density on it

$$E_{surf} = 2\kappa / r = 4 * 10^7 = 2eN_e * 300 / r, \qquad (6)$$

Thus, $N_e = 4 \times 10^{10}$.

9. Acceleration of particles by the electrostatic field

Let us now calculate the resulting ratio Z / A. The proportion for tungsten is

$$6 * 10^{23} \text{ nucleons } - \qquad 184 \text{ g}$$

$$N_{tung \, nucleons} - \qquad 19.35 \text{ g}. \qquad (7)$$

We find from it that in one cubic centimeter of tungsten there are $N_{tung \, nucleons} = 1.16 * 10^{25}$ nucleons. In the wire segment with the diameter $d_{sh} = 6$ μ and length $l_{sh} = 1$ cm there are $3.25 * 10^{18}$ nucleons, and the ratio Z / A for this segment is

$$Z / A = 4 * 10^{10}/3.25 * 10^{18} = 1.23 * 10^{-8}. \qquad (8)$$

Let the accelerating electrostatic field be $E_{acc} = 10$ kV / cm. The equation of motion of particles with a specific (per nucleon) charge $Z / A = 1.23 * 10^{-8}$ can be written as

$$dV / dt = (Z / A) eE_{acc} / M_n, \qquad (9)$$

where M_n is the mass of the nucleon. Assuming that the initial particle velocity is zero, we obtain an expression for the time dependence of the dimensionless velocity $\beta = V / c$, where $c = 3 * 10^5$ km / s is the speed of light in vacuum, in the form

$$\beta = \{(Z / A) eE_{acc}/M_n c^2\} * ct. \qquad (10)$$

Let the final acceleration velocity be $V_{fin} = 1$ km / s, which corresponds to the dimensionless velocity $\beta_{fin} = 3.3 * 10^{-6}$. From equation (10) we can find that $ct = 2.7 * 10^7$. Therefore, the time of acceleration from the initial zero velocity to the final velocity $V_{fin} = 1$ km / sis $t_{acc} = 9 * 10^{-4}$ s.

The acceleration length L_{accl} can be found from the formula of uniformly accelerated motion, which in our case can be written as

$$L_{accl} = \{(Z / A) eE_{acc}/M_n c^2\} * (ct_{acc})^2/2. \tag{11}$$

Substituting numbers into (11), we find that the acceleration length is $L_{accl} = 45$ cm; accordingly, the electrostatic voltage required for the acceleration of particles is U_{acc}

$$U_{acc} = E_{acc} * L_{accl} = 450 \text{ kV}. \tag{12}$$

10. Leakage of electrons

Let us find the number of electrons that left the particle during the acceleration. Given the field strength $E = 40$ MV / cm and the work function $e\varphi = 6.5$ eV, we find from the graph [2] that the leakage current density is $j = 10^{-4}$ A/cm^2.

The charge leakage ΔQ will be:

$$\Delta Q = j * S_{surf} * t_{acc}, \tag{13}$$

where $j = 10^{-4}$ A/cm^2 is the leakage current, $S_{surf} = 1.88 * 10^{-3}$ cm^2 is the lateral surface area of the particle, and $t_{acc} = 9 * 10^{-4}$ s is the acceleration time. Substituting numbers into (13), we obtain $\Delta Q = 10^9$ electrons, which is about 2.5% of the number of electrons planted on the particle.

Gold-coated tungsten wire 5 to 500 μm in diameter has been produced for decades by the Swedish company LUMA [3].

11. Electrodynamic acceleration

A charged tungsten wire segment can be accelerated from the initial velocity $V_{in} = 1$ km / s to the final velocity $V_{fin} = 10$ km / s in a spiral waveguide with a voltage pulse running in its turns.

The final velocity $\beta_{fin} = 3.3 * 10^{-5}$ corresponds to the energy $W_{fin} = Mc^2 \beta^2_{fin} / 2 = 0.54$ eV / nucleon. At the accelerating rate $E_0\sin\varphi_s = 100$ kV / cm this energy can be obtained over the length

$L_{acc} = W_{fin} / [(Z / A) E_0 \sin\varphi_s] = 4.5$ m.

Taking the synchronous phase $\varphi_s = 60^0$, $\sin\varphi_s = 0.87$, we find that the amplitude of the electric field strength in a spiral waveguide must be $E_0 = 115$ kV / cm. It is not necessary to use a sine wave for accelerating a wire segment. Instead, we can use a single sine pulse equal to the half-cycle wave in time and half-length of the wave propagating in the spiral waveguide [4].

The relationship between flux and the electric field strength for a spiral with the space between the spiral and the screen filled with a dielectric is given by the formula [4]

$$P = (c / 8) r_0^2 E_0^2 \beta\varepsilon \{\}, \tag{14}$$

where $c = 3 \times 10^{10}$ cm / s is the speed of light in vacuum, $r_0 = 50$ cm is the initial radius of the spiral, $E_0 = 115$ kV / cm is the field strength on the axis of the spiral, β is the phase velocity, and $\varepsilon = 1280$ is the dielectric constant of the filling between the spiral and the screen; the value of the brackets $\{\}$ for the argument $x = 1$ is approximately 4.

The argument of the modified Bessel functions of the first and second kind, $x = 1 = 2\pi r_0/\beta\lambda_0$, where β is the phase velocity and $\lambda_0 = c/f_0$ is the vacuum wavelength, is optimal [4]. In our case, at the beginning of acceleration $r_0 = 50$ cm, $\beta = 3.3 * 10^{-6}$, $f_0 = 160$Hz, and $T_0 / 2 = 3.12$ ms. Substituting numbers into (14), we obtain

$$P \text{ (W)} = (c / 8) * 2.5 * 10^3 * (1.15)^2 * 10^{10} * 3.3 * 10^{-6} * 1.28 * 10^3 * 4 / (9 * 10^4 * 10^7) = 2.32 \text{ GW}. \tag{15}$$

The length of the slowed-down wave at the beginning of the acceleration is $\lambda_{slow} = \beta\lambda_0 = 3.14$ m, and the spatial length of the accelerating pulse is equal to half the wavelength $\Delta l_{pulse} = 1.57$ m. The base duration of the pulse is $T_0 / 2 = 3.12$ ms. The amplitude of the pulse can be evaluated from the relation

$$U_{pulse} = E_0 * \lambda_{slow}/2\pi = 5.75 \text{ MV}. \tag{16}$$

From the relation $x = 1 = 2\pi r_0/\beta\lambda_0$ it follwos that for each phase velocity of the wave propagating in the spiral waveguide there is its own optimum spiral radius, optimum wavelength, frequency, and pulse duration. As it is quite difficult to cover an order-of-magnitude change in of the phase velocity from

$V_{in} = 1$ km / s to $V_{fin} = 10$ km / s with one pulse duration value, it is reasonable to divide the spiral waveguide into sectionsin which a pulse of optimum duration propagates [4]. Since this pulse defocuses the accelerated particle, there should be focusing elements placed between the sections. In our case, electrostatic quadrupole lenses are most suitable [4].

12. Release of particles into the atmosphere

Particles must be irradiated with an electron beam and then accelerated in a rather high vacuum $P_1 \approx 10^{-9}$ mm Hg, while they are expected to be used under normal atmospheric conditions $P_2 \approx 10^3$ mm Hg, which gives the pressure difference of about 12 orders of magnitude. This pressure gradient can be produced using several buffer cavities, for example, cylindrical chambers separated by end walls with pulse diaphragms built into them. Each cavity must be individually evacuated.

Let us calculate the number of the air particles that penetrate the first cavity, nearest to the atmosphere. Let the radius of the diaphragm be $r_{0d} = 1$ cm and the time for which it remains open be $t_1 = 10^{-4}$ s. Then the linear velocity of the iris diaphragm petals will be $V_{11} = r_{0d} / t_1 = 10^4$ cm / s, which should not be a problem for the operation of the mechanism. The average velocity of the thermal motion of air molecules V can be assumed to be equal to the speed of sound in the air $V = 3 * 10^4$ cm / s; of all possible spatial orientations of the velocity only 1/6 (1/6 is one face of the cube) is directed towards the diaphragm. The number of molecules in a cubic centimeter of the air under normal conditions, the Loschmidt number, of is $\rho_{0L} = 2.7 * 10^{19}$ molecules/cm³. Then the number of molecules coming from the atmosphere to the first buffer cavity while the diaphragm is open is

$$N_0 = (1/6)\, \rho_{0L} * \pi r_{0d}^2 * V_{t1}. \qquad (17)$$

Substituting numerical values into (14), we find that the total number of particles coming from the atmosphere to the first buffer cavity is $N_0 = 4 * 10^{19}$. Let the volume of the first cavity be $V_{01} = 10\,1 = 10^4$ cm³. Then the density of molecules in it after the diaphragm operated will be $n_0 = 4 * 10^{15}$ molecules/cm³.

Particle density (and pressure $p = nkT$) in the first buffer cavity is about 4 orders of magnitude lower than the density of particles in the atmosphere under normal conditions. Thus, at least three cavities are needed to produce the appropriate pressure gradient.

We now consider the dynamics of the particle density in the cavity for the time between the operations of the device. Let the cavity be pumped through the hole with an area $S_1 = 10^3$ cm^2. We assume that all molecules within this area are removed from the volume and that the time between the shots is $t_2 = 10^{-2}$ s, i.e., the operation frequency of the device is F = 100 Hz. The equation describing the particle density reduction during the pumping can be written as

$$dn = -n * (1/6) * S_1 V_t / V_{01}. \tag{18}$$

The solution of this equation can be written as:

$$n = n_0 * \exp[-(1/6) * S_1 V_t / V_{01}] \tag{19}$$

For the pumping time $t = t_2$ the exponent is approximately 5, and thus the pumping reduces the density of molecules in the first buffer cavity by a factor of more than 100. Then, $n = n_0 * 7 * 10^{-3}$, and the density of particles in the first buffer volume before the next shot will be
$n_1 = 4 * 10^{15} * 7 * 10^{-3} = 3 \times 10^{13}$ molecules/cm^3, i.e., 6 orders of magnitude lower than the Loschmidt number $\rho_{0L} = 2.7 * 10^{19}$ molecules/cm^3.

It is evident that before the next shot the cavity can be considered empty.

Figure 1 shows the diagram of the device.

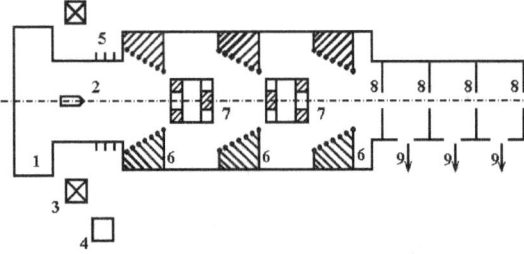

Fig. 1. (1) feeding cassette, (2) particles, (3) pulsed magnetic field coil, (4) electron gun, (5) electrostatic particle acceleration area, (6) - spiral waveguide sections, (7) doublets of quadrupole electrostatic lenses, (8) pulse diaphragms, (9) individual evacuation of buffer cavities.

Application

1. Ballistics

Let us calculate the motion of the particle accelerated by the electrostatic fieldwith air resistance taken into account. The equation of motion of the particle can be written without allowance for the gravitational attraction to the Earth

$$m dV / dt = - \rho C_x S_{tr} V^2 / 2, \qquad (20)$$

where m is the mass of the particle, V is the velocity, $\rho = \rho_0 e^{-z/H0}$ is the barometric formula for the atmospheric density change with altitude, $\rho_0 = 1.3 * 10^{-3}$ g/cm^3 is the air density at the surface of the Earth, $H_0 = 7$ km is the altitude at which the density drops by a factor e. In our case, the particle is a cylinder, S_{tr} is the cross-section of the particle, $S_{tr} = \pi d^2 / 4 = 2.8 * 10^{-7}$ cm^2, and C_x is the aerodynamic coefficient.

The aerodynamic coefficient, or the drag coefficient, is a dimensionless quantity that takes into account the "quality" of the particle form

$$C_x = F_x / (½) \rho_0 V_0^2 S_{tr}. \qquad (21)$$

Equation (20) can be written as

$$V (t) = V_0 / [1 + \rho C_x V_0 * S_{tr} * t/2m]. \qquad (22)$$

To calculate the particle velocity variation in time, it is necessary to find the aerodynamic coefficient C_x.

2. Calculation of the particle drag coefficient in the air

We assume that the particle has the shape of a cylindrical rod with a conical head. When a nitrogen molecule hits the cone, the change in the longitudinal velocity of the molecules is

$$\Delta V_x = V_x * \Theta_h^2 / 2, \qquad (23)$$

where Θ_h is the cone angle. The momentum transferred by the gas molecules to the macroparticle is

$$p = mV = \rho V_x S_{tr} t * \Delta V_x. \qquad (24)$$

The change in the momentum per unit time is a force, namely, a drag force.

$$F_{x1} = (\tfrac{1}{2}) \rho V_x S_{tr} * V_x * \Theta_h^2. \qquad (25)$$

Dividing Fx1 by $(\tfrac{1}{2}) \rho V_x^2 S_{tr}$, we obtain the drag coefficient for a pointed cone at the mirror reflection of molecules from the cone (the Newton formula)

$$C_{x\,air} = \Theta_h^2. \qquad (26)$$

Let the length of the conical part of the particles be $l_{cone} = 3$ mm. This means that the cone angle is $\Theta_h = 2 * 10^{-3}$ and $C_{x\,air} = 4 * 10^{-6}$.

3. Passage of particles through the atmosphere

Let us calculate the passage of the particle with a drag coefficient $C_{x\,air} = 4 * 10^{-6}$ through the Earth's atmosphere. The decrease in the velocity with time due to air resistance is described by (20). We choose the following parameters: $\rho = \rho_0 e^{-z/H0}$, where $H_0 = 7$ km, $C_{x\,air} = 4 * 10^{-6}$, $S_{tr} = 3 * 10^{-7}$ cm^2, $V_{fin} = 10^6$ cm / s, $m = 6 * 10^{-6}$ g, and $\rho_0 = 1.3 * 10^{-3}$ g/cm^3. For the first few seconds of flight the decrease in the velocity is very small.

Further, as the altitude increases, the contribution of this term will decrease exponentially. Thus, for the altitude $z = 2$ km the factor $e^{z/H0}$ is 0.75, for $z = 4$ km it is 0.57 etc. This means that the velocity loss of the particle during its ascent and descent can be neglected and it can be considered as flying in vacuum.

4. Particle penetration depth in aluminum

The depth to which a tungsten wire segment moving at a velocity of $V_0 = 10$ km / s penetrates in aluminum can be estimated in the following way. Since the processes in question are fast, it is necessary to consider only interaction with the atoms that are on the way of the wire segment.

Let us find the energy to be spent for heating one centimeter of aluminum with the cross sectional area equal to the cross-section of the tungsten wire

$S_{tr} = 2.8 * 10^{-7}$ cm^2 from room temperature to the melting point. Then we should add the solid–liquid phase transition energy to this energy. The fluid should be heated to the evaporation temperature and the liquid–vapor phase transition energy should be taken into account.

The total expenditure of energy is $\Delta W_{1\,cm} = 10^{-2}$ J / cm.

The kinetic energy of the tungsten wire segment with a diameter of $d_{sh} = 6$ μ, length $l_{sh} = 1$ cm, and velocity $V_0 = 10$ km / s is $W_{sh} = 0.3$ J. the analysis of collisions of meteorites with solids allows the conclusion that at these velocities about 40% of the kinetic energy of meteorites is spent for heating and evaporation of the target material.
We assume that out of the entire kinetic energy of the tungsten wire segment $W_{sh} = 0.3$ J, the energy spent for evaporation of aluminum is $W_{sh1} = 0.1$ J.

Now, dividing W_{sh1} by the specific energy loss $\Delta W_{1\,cm}$, we find the depth of penetration of the segment in aluminum:
$L_{Al} = W_{sh1} / \Delta W_{1\,cm} = 10$ cm.

5. Drag-affected penetration depth of particles in water

Let us find the drag force acting on the particle moving in water. Unlike the case in the air, reflection of molecules from the particle in water should be considered as being diffuse, which leads to the drag coefficient $C_{x\,w} \approx 1$.
The time dependence of the veclocity reduction will be described by (23). We denote the particle stopping velocity in water by Vs. Let Vs be 0.1 km / s. From (23) we obtain the particle slowing-down length

$$l_s = \int_0^{t_0} V(t)\, dt = [2m / (\rho_w * C_{x\,w} * S_{tr})] * \ln(1 + V_0/V_s), \qquad (27)$$

where V_0 is the velocity at which the particle enters water, $\rho_w = 1$ g/cm^3 is the density of water, $C_{x\,w} = 1$, m is the mass of the particle, and $V_s = 0.1$ km / s. Thus, the particle slowing-down length in water is

$$l_{s1} = \{2m / (\rho_w * C_{x\,w} * S_{tr})\} * \ln(1 + V_0/V_s), \qquad (28)$$

The coefficient of the factor $\ln(V_0/V_s - 1)$ has the dimension of length and is 40 cm for the parameters chosen above. In our case, $V_0/V_s = 10^2$, and relation (28) yields the drag-affected particle penetration depth in water $l_{s1} = 180$ cm.

Contributions to particle slowing-down in water will also come from the other processes, e.g., viscosity of water.

6. Penetration depth of particles in water as affected by water viscosity

The equation of motion of the particle entering water can be written as

$$m dV_x / dt = -F_{x2} \qquad (29)$$

The drag force due to viscosity is

$$F_{x2} = \sigma S = \eta \, (dV / dr) * 2\pi r_s h \, l_{sh} = \eta \, (V / \delta) * 2\pi r_{sl}, \qquad (30)$$

where $\eta = 10^{-2}$ Poise is water viscosity, $r_{sh} = 3 * 10^{-4}$ cm is the radius of the particle, $l_{sh} = 1$ cm is the particle length, dV / dr is the radial gradient of the longitudinal water velocity near the particle, and δ is the characteristic length of the velocity variation in radius.

Equation (29) with the viscosity drag force in the form of (30) can be rearranged in the form

$$dV / V = - (\eta / m\delta) * 2\pi r_{sh} \, l_{sh} * dt, \qquad (31)$$

which has the solution

$$V = V_0 \exp [- (\eta / m\delta) * 2\pi r_{sh} l_{sh} * t]. \qquad (32)$$

Now we can find the path traveled by the macro particle to a stop in water

$$l_{water} = \int_0^\infty V dt = V_0 \, (m\delta / \eta * 2\pi r_{sh} \, l_{sh}). \qquad (33)$$

To find the particle penetration depth in water, we need to find δ, the characteristic length of the ramp near the water velocity variation in radius near the particle.

Let us find it from the Navier equation

$$\partial V_x / \partial t + V_x \, \partial V_x / \partial x = (\eta / \rho r) \, d / dr \, (r dV_x / dr) + \partial^2 V_x / \partial x^2. \qquad (34)$$

18

First we will see in which case the term $V_x \, \partial V_x / \partial x$ is much greater than $(\eta / \rho) \, \partial^2 V_x / \partial x^2$. Let $\partial x = l_{ch}$ be the characteristic length over which the velocity varies. Then $V_x \, \partial V_x / \partial x = V_x^2 / l_{ch}$, $(\eta / \rho) \, \partial^2 V_x / \partial x^2 = (\eta / \rho) \, V_x / l_{ch}^2$. The condition $V_x \, \partial V_x / \partial x \gg (\eta / \rho) \, \partial^2 V_x / \partial x^2$ is satisfied if $\rho V_x l_{ch} / \eta = Re \gg 1$, where Re is the Reynolds number and the relation $Re \gg 1$ is certainly satisfied in our case.

Now we rearrange (34) in the form

$$\partial V_x / \partial t + V_x \, \partial V_x / \partial x = dV_x / dt = (\eta / \rho r) \, d / dr \, (r dV_x / dr), \quad (35)$$

where we replaced $\partial / \partial t + V_x \, \partial / \partial x$, the partial derivative with respect to time and coordinate by d / dt, the total time derivative.

Substituting $dV_x / dt = -(\eta / m) * 2\pi r_{sh} l_{sh} * (dV_x / dr)$ found from (33) into the Navier equation, we obtain the equation

$$-(\eta/m)*2\pi r_{sh} l_{sh} *(dV_x/dr)= (\eta/\rho r)d/dr(r dV_x/dr), \quad (36)$$

which, after cancellation of η (the water viscosity) and transpositions, takes the form

$$2\pi r_{sh} l_{sh}\rho / m * (r dV_x / dr) = -d/dr \, (r dV_x / dr). \quad (37)$$

Replacing $r dV_x / dr$ by y, we get:

$$d \, (\ln y) = - \, (2\pi r_{sh} \, sh \, \rho / m) \, dt, \quad (38)$$

or

$$dV_x / dr = (C_{1v} / r) \exp (-2\pi r_{sh} \, l_{sh} \, \rho r / m). \quad (39)$$

To find the integration constant C_{1v}, we integrate (39) again and obtain

$$V = C_{1v} * Ei \, (2\pi r_{sh} \, l_{sh} \, \rho r / m) + C_{2v}, \quad (40)$$

where $Ei \, (z) = \int_0^\infty \exp (-t) \, dt / t$ - expint (z) is the exponential integral. Here we made a replacement $2\pi r_{sh} \, l_{sh} \, \rho r / m = t$; the limits of integration with respect to r are from $r = r_{sh}$ to infinity, and the lower limit of integration with respect to t is accordingly $t = 2\pi r^2_{sh} l_{sh} \rho / m = 2\rho_{water}/\rho_{tungsten} = 0.1$.

The function Ei (z) cannot be expressed in terms of elementary functions; it can be approximated as [5]

$$(\tfrac{1}{2}) e^{-z}\ln (1 + 2 / z) < Ei (z) < e^{-z}\ln (1 + 1 / z), z > 0, \qquad (41)$$

Equation (40) describes the radial dependence of the velocity of the medium as the particle moves in it at a velocity V. At large distances from the particle $V = 0$, and thus $C_{2v} = 0$. The boundary condition on the surface of a body is usually taken to be the condition of "sticking" of the medium particles to the body, i.e, when $r = r_{sh}$, the velocity of the medium is $V = V_0$, from which we can find the constant C_{1v}

$$V_0 = C_{1v} * Ei (0.1) \text{ and}$$

$$C_{1v} = V_0/Ei (0.1). \qquad (42)$$

Expression (32) for the velocity gradient at the surface of particles can now be written as

$$dV_x / dr = V_0 / [r_{sh} Ei (0.1) * \exp (0.1)]. \qquad (43)$$

Thus, the characteristic amount of the radial velocity variation δ is

$$\delta = r_{sh} * Ei (0.1) * \exp (0.1), \qquad (44)$$

and, according to formula (33), the particle path length particles in water is

$$l_{water} = \int_0^\infty V dt = V_0 (m\delta / \eta * 2\pi r_{sh} l_{sh}) = V_0 [m * Ei (0.1) * \exp (0.1) / 2\pi\eta l]. \; (45)$$

Finally,

$$l_{water} = (V_0 m/2\pi\eta l_{sh}) * Ei (0.1) * \exp (0.1). \qquad (46)$$

Substituting $V_0 = 10 \text{ km} / \text{s}$, $m = 6 * 10^{-6}$ g, $\eta = 10^{-2}$ Poise, $l_{sh} = 1$cm, and $Ei (0.1) = 2.6$, $\exp (0.1) = 1.1$ into (46), we obtain

$$l_{water} = 230 \text{ cm}. \qquad (47)$$

To find the penetration depth in water as affected by two processes, viscous drag and friction, we should be add the inverse values and find the penetration depth L_{water} by the formula

$$L_{water} = l_{s1} * l_{water} / (l_{s1} + l_{water}) = 1m. \qquad (48)$$

Thus, the penetration depth of the tungsten wire segment in water is approximately $L_{water} = 1m$.

Conclusion

Even a weakly magnetized particle will rotate in the magnetic field of the Earth like an ordinary compass needle. If the particle turns relative to the velocity vector, all aerodynamics and penetration depth formulas will be incorrect, which means that we must look for other ways to orient macro particles relative to the axis of acceleration.

If we apply four sectors of a material with different resonant absorption frequencies to the macro particle, the particle motion can be controlled by evaporation of the material from the corresponding sector and transverse ablative acceleration of the particle. Considering that the velocity of the evaporated material is $V_{ab} = 0.1$ km / s and the mass of the evaporated material is $m_{ab} = 0.1$ m (0.1 of the wire segment mass), we find that in this way we can get angles of deviation from the original direction on the order of $\theta_\perp = 10^{-3}$.

References

1. http://ru.wikipedia.org/wiki/Пушка_Гаусса
2. Tables of Physical Data. The Handbook, edited by I.K. Kikoin, Moscow, Atomizdat, 1976
3. http://www.luma-metall.se/
4. S.N. Dolya, K.A. Reshetnikova, About the electrodynamic acceleration of macroscopic particles, Communication JINR P9-2009-110, Dubna, 2009, http://www1.jinr.ru/Preprints/2009/110(P9-2009-110).pdf
http://arxiv.org/ftp/arxiv/papers/0908/0908.0795.pdf
5. http://en.wikipedia.org/wiki/Exponential_integral

Electromagnetic acceleration of electrically charged bodies

Acceleration of electrically charged bodies is carried out by the electric field running via the spiral structure of the electric pulse. The accelerated particles have a cylindrical shape with a diameter of cylinder $d_{sh} = 2$ mm, a length of the conical part $l_{cone} = 13$ mm and the total length $l_{sh} = 300$ mm. Pre-acceleration of the cylinder up to speed $V_{in} = 1$ km / s is performed by gas-dynamic. The pulse with the voltage amplitude $U_{acc} = 2$ MV and the power $P = 300$ MW goes into the spiral waveguide synchronously with the rod injected onto it. The rod is accelerated by the traveling pulse in the longitudinal direction up to the finite velocity $V_{fin} = 6$ km / s for length $L_{acc} = 300$ m.

Introduction

There is a known [1] method of magnetic dipole acceleration by sequential current turns, which are switched on one by one creating the current pulse moving in the space. Magnetic dipoles in this process are accelerated by the magnetic field gradient in the space. In principle, using a large number of turns allows one to achieve a high finite speed of the magnetic dipoles.

Such a multiple-section accelerator consisting of a sequence of coils and capacitors can be considered as a line with the lumped parameters. If you move from the line with lumped parameters to the line with distributed parameters, you will obtain a casual coaxial cable, which has the central wire rolled into a spiral or on a spiral waveguide.

In such a cable in a wide range of frequencies there is no dispersion, i.e. there is no dependence of the phase velocity on the wave frequency. The phase velocity in this cable coincides with the group velocity of the wave. The wave propagation velocity (pulse), V, in such a cable is defined by the tightness of winding of the central conductor in a spiral as well as by the dielectric properties of the medium which fills the cable. This ratio is called the dispersion equation and looks as follows:

$$\beta = tg\Psi/\varepsilon^{1/2}, \qquad (1)$$

$\beta = V / c$, V - velocity of the pulse propagation via the cable,
$c = 3 * 10^{10}$ cm / s - the speed of propagation of electromagnetic waves in the vacuum, $tg\ \Psi = h/2\pi r_0$, h -the winding step of the spiral, r_0 - the radius of the spiral winding, ε - relative dielectric constant of the medium filling the cable. The wave as if runs over the circle spiral around $2\pi r_0$ while moving at a small distance h along the axis of the spiral. Further the wave additionally slows

down due to the dielectric properties of the medium defined by the value of ε.

To speed up the rod by the pulse running in the cable, the area inside the spiral must be released from the dielectric. Thus, the speed of the pulse in this cable will be slightly increased, [2]:

$$\beta = \sqrt{2} * \text{tg } \Psi / \varepsilon^{1/2}. \qquad (2)$$

Running on the line with lumped or distributed parameters the pulse contains not only the gradient of the magnetic field which accelerates the magnetic dipoles, but also the electric field E_{zw}, which can accelerate charged bodies.

1. The parameters of the body being accelerated

We will consider the acceleration of the rod with a conical head which is electrically charged.

Acceleration of macro particles in a spiral waveguide is well-known, [2]. For this acceleration it is required that the initial velocity of the rod and the phase velocity are approximately the same. When we accelerate the rod, the phase velocity in the spiral waveguide should be increased for the rod to be all the time in one the same phase of the wave, which is called a synchronous phase. To increase the phase velocity of the wave in the waveguide is possible by increasing the winding step of the spiral or decreasing its radius, or doing the both simultaneously, [2].

Let the diameter of the rod be equal to: $d_{sh} = 2$ mm, length $l_{sh} = 300$ mm. Then the cross-section area of the rod is: $S_{tr} = \pi d_{sh}^2 / 4 = 3.14 * 10^{-2}$ cm^2, the volume of the rod: $V_{sh} \approx 1$ cm^3. The weight of the rod in the case of the average density of the rod $\rho_{aver} = 5$ g/cm^3, is equal to $m_{sh} = 5$ g.

1.1. The ratio Z / A

We assume the average atomic mass of the rod to be equal to $A_{sh} = 30$. We find the number of nucleons in the rod from the following proportion:

$$6 * 10^{23} \text{ -------- } 30 \text{ g}$$
$$x \text{ ---------- } 5 \text{ g,}$$

where $x = 10^{23}$ atoms or $A_{sh} = 3 * 10^{24}$ nucleons.

We take the surface tension of the electric field on the rod to be: $E_{surf} = 3 * 10^7\,V/cm$. The formula for the surface tension of the electric field for the cylinder is the following:

$$E_{surf} = 2\kappa / r, \tag{3}$$

we find the charge density per the length unit of the rod:

$$\kappa = E_{surf}\,r/2e = (3 * 10^7 * 0.1)/(5 * 10^{-10} * 300 * 2) = 10^{13}, \tag{4}$$

from where you can find

$$N_e = (\kappa / e) * l_{sh} = 3 * 10^{14}. \tag{5}$$

Thus, if the "put" $N_e = 3 * 10^{14}$ electrons on the rod, the surface tension of the field will turn out to be: $E_{surf} = 3 * 10^7\,V/cm$.

Now when we know the total number of the electrons "placed" on the rod: $N_e = 3 * 10^{14}$, and the number of nucleons on it: $A_{sh} = 3 * 10^{24}$, it is possible to find the ratio of the charge to the mass for the rod: $Z/A = N_e/A = = 3 * 10^{14}/3 * 10^{24} = 10^{-10}$.

1. 2. Irradiation of the rod with the electron beam

To accelerate a cylindrical rod with a cone head in a spiral waveguide, it is necessary to charge it electrically. To give the electric charge to the rod is possible by irradiating it with the electron beam; so that the electrons irradiating the rod would obligatory remain on it. Then the electric charge of the rod will increase proportionally to the current of the electron beam and the duration of exposure of the rod. Suppose that the current of the electron beam irradiating the rod is equal to $I_{beam} = 5\,A$, and the current pulse duration is $\tau_{beam} = 10\,\mu s$. Then the total number of electrons in the current pulse is exactly equal to $N_e = I_{beam} * \tau_{beam} / e = 3 * 10^{14}$ electrons.

1.3. Irradiation of the rod with the electron beam. The electron energy

Let the cylindrical rod accelerated by the gas-dynamic method up to speed $V_{in} = 1\,km/s$ be irradiated with the electron beam obtained from an external source. We assume the surface tension of the electric field to be equal to

E_{surf} = 30 MV / cm. Then, for the diameter of the cylinder d_{sh} = 2 mm, we find that the minimum energy of the electrons which can overcome the Coulomb repulsion of the electrons previously placed on the rod should be:
$W_e > eE_{surf} * d_{sh} / 2 = 3$ MeV.

1. 4. Irradiation of the rod with the electron beam. The path length of the electrons in the rod

Electrons with the energy of 3 MeV have the range path in aluminum approximately equal to 1 g/cm², [3], page 957. Assuming the density of aluminum to be equal to: $\rho_{Al} = 2.7$ g/cm³, we find that the extrapolated path of electrons in aluminum is: $l_{Al} \approx 4$ mm. Since the average density of the material chosen for the cylinder ρ_{aver}= 5 g/cm³, that is by about twice more than the density of aluminum, the path length of electrons of the energy of 3 MeV in the rod will be approximately equal to 2 mm.

Evidently, it is necessary to gradually increase the energy of the electrons in the process of irradiation. It is needed that while "setting" the electrons on the rod, the electrons emitted later, on the one hand, would have a sufficiently high energy to overcome the Coulomb repulsion of the electrons being already on the rod, and, on the other hand, the electron energy must not be too high because it is necessary to have the length of the electron path in the material of the rod to be much less than its diameter.

In this energy range the length of the electron passing in the rod material linearly increases with the energy, for example, of the electrons with energies W_e = 300 keV, having the length of electron passing equal to 0.2 mm. They cannot cross the rod diameter of 2 mm. They will lose their energy for ionization of the matter, and will be placed on the rod.

1. 5. Irradiation of the rod with the electron beam. Field electron emission

To "plant" several charges on the rod is not a problem, but when there are many electrons on the rod, they will start to leak out from it due to the field emission. Let the field strength for the field emission be $E_{surf} = 3 * 10^7$ V / cm. When there are enough "planted" electrons on the rod, to plant the next portion, it is necessary to overcome the repulsion of those electrons which are already sitting there. This means that the energy of the electrons, which we would like to put on the rod, should be large enough so that they can overcome this Coulomb barrier, reach the rod and stay on it.

"Planting" a large electric charge would be interfered by the field emission. A part of the charge due to the tunnel effect will continuously leak out from the rod.

1.6. Surface covering with platinum and oxygen passivation of the cylinder

To create a surface barrier for the electrons "having placed" on the rod, it is needed to increase the energy yield of the electrons from the rod. The largest energy yield belongs to platinum passivated by oxygen,
$e\varphi = 6.56$ eV, [3], page 445. Planted on the rod the charge will leak out from it by the field emission according to the formula [3], page 444:

$$j = e^2E^2/(8\pi h\varphi)*\exp\{\ [-(8\pi/3)(2m)^{1/2}/h]*[(e\varphi)^{3/2}/(eE)*\theta(y)]\}, \quad (6)$$

where $\theta(y)$ is the Nordheim function. The argument of this function is a relative reduction of the energy yield by the external electric field according to Schottky's law.

1.7. Leakage of electrons

Let us find the number of electrons which will leave the rod during acceleration. For the field tension $E = 30$ MV / cm and the energy yield $e\varphi = 6.5$ eV from the graph, [3], page 461, we find that the leakage current density is: $j = 10^{-9}$ A/cm^2.

Leakage of charge ΔQ will be:

$$\Delta Q = j * S_{surf} * t_{acc}, \quad (7)$$

where $j = 10^{-9}$ A/cm^2 - leakage density current, $S_{surf} \approx 20$ cm^2 – the total surface of the rod.

The acceleration can be determined from the following formula:

$$t_{acc} = L_{acc} / V_{aver}, \quad (8)$$

where $L_{acc} = 300$ m - length of acceleration, $V_{aver} = 3$ km / s - the average speed over the length of the acceleration. Calculating the time of the acceleration from formula (8) we find it to be equal to: $t_{acc} = 0.1$ s.

Substituting numbers into the formula (7) we find that $\Delta N_e = 10^{10}$ electrons and it is $3 * 10^{-5}$ - the number of electrons, which were planted on the rod.

2. The acceleration length

The acceleration rate of the charge in the electric field can be written as follows:

$$\Delta W = (Z / A) e E_{zw}, \qquad (9)$$

and, for the tension of the wave $E_{zw} = 70$ kV / cm, the rate of the energy gain will be: $\Delta W = 7 * 10^{-4}$ eV / (m * nucleon), so that the required increase of energy $W_{fin} = 0.2$ eV / nucleon will be reached on the length:

$$L_{acc} = W_{fin} / \Delta W = 300 \text{ m}. \qquad (10)$$

3. Selection of parameters of the spiral waveguide

The initial velocity of the rod in a spiral $\beta_{sh\ in}$ expressed in terms of the velocity of light $\beta_{sh\ in} = V_{sh\ in} / c$, where $c = 3 * 10^{10}$ cm / s, the velocity of light in vacuum is equal to $\beta_{sh\ in} = 3.3 * 10^{-6}$, finite $\beta_{sh\ fin} = 2 * 10^{-5}$. The spiral is assumed to consist of several sections, so that within each section to select the optimal acceleration rate. The wavelength of the acceleration can be determined from the condition: $x = 2\pi r_0 / (\beta_{ph} * \lambda_0) = 1$, where x - a dimensionless parameter which is the argument of the modified Bessel functions, r_0 - the radius of the spiral, β_{ph} - phase velocity, λ_0 - wavelength acceleration in the vacuum, $\lambda_0 = c/f_0$, f_0 – acceleration frequency .

Choosing the initial radius r0 in the spiral equal to $r_{0\ in} = 20$ cm, $\varepsilon = 1280$ - the dielectric constant of the medium located in the area between the coil and the screen, we find: $\lambda_0 = 3.8 * 10^7$ cm, $f_0 = 790$ Hz. Thus, the slowdown wavelength for the start of acceleration is equal to: $\lambda_{slow} = \beta\lambda_0 = 1.25$ m.

3.1. Parameters of the spiral

In order to obtain the required field intensity E_0 in a spiral waveguide, it is required to introduce the power, defined by the formula, [2]:

$$P = (c / 8) * E_0^2 * r_0^2 * \beta_{ph} * \{\}, \tag{11}$$

where P – the high frequency power introduced into the spiral waveguide, r_0 - the radius of the spiral, β_{ph} - phase velocity of the wave, which is determined from the dispersion equation. The curly bracket in (11) is equal to:

$$\{\} = \{(1+I_0K_1/I_1K_0)(I_1^2-I_0I_2) + \varepsilon \, (I_0/K_0)^2(1+I_1K_0/I_0K_1)(K_0K_2-K_1^2)\}, \tag{12}$$

where I_0, I_1, I_2 are the modified Bessel functions of the first kind, K_0, K_1, K_2 - the modified Bessel functions of the second kind. The first term in the curly bracket corresponds to the flux propagating inside the spiral; the second term corresponds to the flux traveling outside the spiral. Since the space between the spiral and the screen is filled with a dielectric, before it there is the second term factor ε, [2].

In this case, to slow down the electromagnetic wave till the velocity of sound, it is required to use geometrical properties of the structure (spiral small step) as well as the properties of the medium. That is why we have chosen the relative permittivity $\varepsilon = 1280$.

Thus, the flow of high frequency power propagating outside the spiral is more than 10^3 times greater than the power propagating inside the spiral. Therefore, the first term inside the curly bracket can be neglected compared to the second one. The very meaning of the bracket for the argument $x = 1$ is approximately equal to: $\{\} \approx 4\varepsilon$.

In accelerators the synchronous phase is selected on the front slope of the pulse, so that the electric field accelerating the particle is always less than the amplitude value. Let us choose a synchronous phase to be equal to: $\varphi_s = 45^0$, $\sin\varphi_s = 0.7$, $E_{zw} = E_0\sin\varphi_s$. Thus, the amplitude of the wave which accelerates the cylindrical rod should be equal to the following:

$$E_0 = E_{zw} / \sin\varphi_s = 100 \text{ kV} / \text{cm}. \tag{13}$$

Then, the wave power, expressed by the formula (11) in Watts, is equal to:

$$P \, (W) = 3*10^{10}*10^{10}*4*10^2*3.3*10^{-6}*1.28*10^3*4/(8*9*10^4*10^7) =$$
$$= 300 \text{ MW}. \tag{14}$$

3.2. The transition from a sine wave to a single pulse

This power can be achieved by using the pulse technique.
We expand the sinusoidal pulse, [2], the corresponding half-wave
$E_{pulse} = E_{0pulse}\sin(2\pi/T_0) t$, $2\pi/T_0 = \omega_0$, $\omega_0 = 2\pi f_0$ in a Fourier series:

$$f_1(\omega) = (2/\pi)^{1/2} \int_0^{T_0/2} \sin\omega_0 t * \sin\omega t dt. \qquad (15)$$

The pulse spectrum is narrow and covers the frequency range from
0 to $2\omega_0$. Since the spiral waveguide dispersion (dependence of the phase velocity on the frequency) is weak, it can be expected that the full range of frequencies from 0 to $2\omega_0$ will propagate approximately with the same phase velocity. As a result, the half-wave sinusoidal pulse will spread out in the space and becomes wider by several times only due to the increase of the phase velocity of the wave. In this case the spiral waveguide is necessary to match with a supply feeder in the following frequency range: $\Delta f \approx \omega_0/2\pi$.

We introduce the concept of pulse amplitude $\tilde{U}$, associated with the field tension at the axis of the spiral E_0 by the following ratio, [2]:

$$\tilde{U}_{pulse} = E_{0pulse}\lambda_{slow}/2\pi, \quad \lambda_{slow} = \beta\lambda_0, \quad \lambda_0 = c/f_0. \qquad (16)$$

Selection of wavelength $\lambda_0 = 3.8 * 10^7$ cm means that we have chosen the duration of the acceleration of the rod equal to ($f_0 = c/\lambda_0 = 790$ Hz), $\tau_{pulse} = 1/(2f_0) = 630$ μs. The amplitude of the voltage pulse will be equal to: $\tilde{U} = E_0\lambda_{slow}/2\pi = 2$ MV. Table 1 summarizes the main parameters of the accelerator.

Table1. The parameters of the accelerator

$Z/A = 10^{-10}$, insulator outside the spiral, wave power, P	P = 300 MW $\mu = 1$, $\varepsilon = 1280$
Speed, the initial - finite, β_{ph}	$\beta_{ph} = 3.3*10^{-6} - 2*10^{-5}$
Initial radius of the spiral, r_0	$r_0 = 20$ cm
Frequency of the wave, f_0,	$f_0 = 790$ Hz
Tension of the electric field E_0	$E_0 = 100$ kV/cm
Length of the accelerator, L_{acc}	$L_{acc} = 300$ m
Pulse duration, τ	$\tau = 630$ μs
Amplitude of voltage $\tilde{U}_a$	$\tilde{U}_a = 2$ MV

29

3. 3. *The capture of particles in the acceleration mode. Admission*

We calculate the required accuracy to match the initial phase of the accelerating wave (pulse) with a synchronous phase. The theory of particle capture in a traveling wave gives $\Delta\varphi = 3\varphi_s$, $(+ \varphi_s - 2\varphi_s)$, [4]. In our case it means the following: $T_0 / 4$ correspond to the duration of 316 µs or 90^0 degrees, and a one-degree phase corresponds to the time interval of approximately 3 µs. In linear accelerators the buncher gives the phase width of the bunch $\pm 15^0$. To avoid large phase oscillations, it is required that the timing accuracy of synchronization of the rod with the accelerating pulse would be equal to: $\Delta\tau = \pm 15 * 3$ µs $= \pm 45$ µs. This timing precision seems to be quite achievable for the gunpowder start which is a preliminary gas-dynamic acceleration of the rod.

Now let us calculate the required accuracy of the coincidence of the wave phase velocity with the initial rate of the rod. We introduce value $g = (p-p_s) / p_s$ - the relative difference between the pulses, [4]. In the non-relativistic case – it is just the relative velocity dispersion of $g = (V-V_s) / V_s$. The vertical scale of the separatrix is calculated by the following formula, [4]:

$$g_{max} = \pm 2 \, [(W_\lambda ctg\varphi_s/2\pi\beta_s) * (1 - \varphi_s / ctg\varphi_s)]^{1/2}, \qquad (17)$$

wherein: $\varphi_s = 45^0 = \pi / 4$, $ctg\varphi_s = 1$, $[1 - \varphi_s / ctg\varphi_s]^{1/2} = 0.46$, $2 * 0.46 = 0.9$
$W_\lambda = (Z / A) \, eE_0\lambda_0 sin\varphi_s/Mc^2$.

Let us determine the value of $W_\lambda = (Z / A)* eE_0\lambda_0 sin\varphi_s/Mc^2$, which is the relative set rod energy at wavelength λ_0 in vacuum. In our case $\lambda_0 = c/f_0 = 3.8 * 10^7$ cm, $sin\varphi_s = 0.7$, $Mc^2 = 1$ GeV, $W_\lambda = 2.66 * 10^{-6}$. Substituting numerical values, we get $g = (V_{in}-V_s) / V_s = \Delta V / V_s$, and, finally,

$$\Delta V / V_s = \pm [2.66 * 10^{-6} / (6.28 * 3.3 * 10^{-6})]^{1/2} * 0.9 = \pm 0.11.$$

Thus, the accessible mismatch of the rod initial velocity with the pulse velocity is of the order of $\Delta V / V_s = \pm 11\%$. For the initial rate of the rod $V_{in} = 1$ km / s, the mismatch accuracy of the velocity deviation is $\Delta V < 100$ m / s.

4. Radial movement

It is well known, [4], that in the azimuthal - symmetric wave the phase stability region corresponds to the radial defocusing. In this case, you cannot simultaneously obtain both the radial and phase stabilities. Under the conditions of phase stability for radial focusing it is required to use the external field. In this phase region the radial component of the electric field of the wave is directed to the increasing radius, i.e. it radially accelerates the rod.

In this velocity region of the rods – the "hypersonic" velocity region where they are by hundreds of thousand times smaller than the speed of light, the focusing by magnetic quadrupole lenses is not efficient. In this case the most suitable focusing is by using the electrostatic quadrupole lenses. These lenses focus the particles in one plane and defocus them in the other one. Collected into a doublet, two lenses of this sort give the focusing effect. The accelerator is divided into separate sections and the focusing doublets are placed between the accelerator sections.

5. Operation of the device

Fig. 1 shows a diagram of the apparatus.

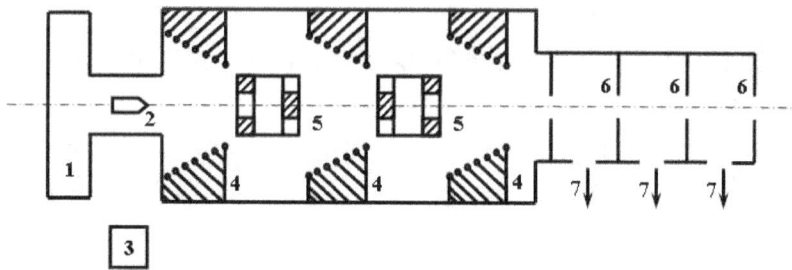

Fig. 1

The apparatus operates as follows. Inside the barrel there is the gun – (1), (2) - a cylindrical rod with a sharp conic head, which is accelerated till the speed corresponding to the speed of injection in a spiral waveguide till $V_{in} = 1$ km / s. From the linear accelerator (3) to the rod which is irradiated by the beam of electrons with energy $eE = 3$ MeV, the total number of electrons is planted on the rod $N_e = 3 * 10^{14}$. Therefore, the electric field tension obtained at the surface of the cylinder is equal to

31

$E_{surf} = 3*10^7$ V / cm, the potential of the cylinder - $e\Phi = 3$ MeV, the ratio of the planted charge to the mass is $Z / A = 10^{-10}$. The electric field potential of the high current pulse with voltage $\tilde{U}_a = 2$ MV, is propagating via sections (4) of the spiral waveguide of the total length of $L_{acc} = 300$ meters. The rod is accelerated to the finite speed $V_{fin} = 6$ km / s. The electrostatic quadrupole lens doublets (5) are placed between the sections and they focus the rods in the transverse direction. The rods are released into the atmosphere through a series of buffer volumes (6). Each buffer volume has individual pumping (7).

Conclusion

The flight parameters of the rod are represented in the Table 2, where they are given as a function of time in the first column. The second column shows the vertical velocity of the rod, the third one - the horizontal velocity of the rod, the fourth column shows the achieved altitude of the rod, the fifth one represents the density of the atmosphere at this altitude.

Table 2. The flight parameters depending on the time, for the case of $C_x, C_y = 2.5*10^{-2}$.

t, s	V_x, km/s	V_y, km /s	Y, km	ρ_{air}, g/cm^3
0	6	0	0	$1.3*10^{-3}$
10	3.72	3.67	18	$6*10^{-5}$

The time of flight up to the maximum altitude is equal to $\tau_{max} = V_y / g = 367$ s, where $g = 10^{-2}$ km/s^2 (gravity acceleration range). The distance of the flight is $S = V_x * 2\tau_{max} = 2700$ km, the maximum altitude is $Y = V^2_y/2g = 670$ km.

Literature
1. http://ru.wikipedia.org/wiki/Пушка_Гаусса
2. S.N. Dolya, K.A. Reshetnikova, 'On the electrodynamics' acceleration of macroscopic particles, Communication JINR P9-2009-110, Dubna, 2009, http://www1.jinr.ru/Preprints/2009/110(P9-2009-110).pdf http://arxiv.org/ftp/arxiv/papers/0908/0908.0795.pdf
3. Tables of physical quantities. Directory Ed. I. K. Kikoin, Moscow, Atomizdat, 1976
4. I. M. Kapchinsky, Particle dynamics in linear resonance accelerators, Moscow, Atomizdat, 1976

II. Acceleration of the magnetic and electric dipoles

Acceleration of magnetic dipoles by the sequence of current turns

Acceleration of magnetic dipoles is carried out by the running gradient of the magnetic field formed while sequent switching on the current turns. Magnetic dipoles, with a diameter of $d_{sh} = 60$ mm and full length $l_{tot} = 1$m, are pre-accelerated by using the gas-dynamic method to speed $V_{in} = 1$ km / s, corresponding to the injection rate into the main accelerator. To prevent the turning of the dipoles by 180 degrees in the field of the accelerating pulse and focus them, the magnetic dipoles are accelerated inside the titanium tube. The magnetic dipoles have mass $m = 2$ kg and acquire the finite speed $V_{fin} = 5$ km / s on the acceleration length $L_{acc} = 300$ m.

Introduction

There is a known [1], method to accelerate the magnetic dipoles by the current pulse moving in the space. Using a large number of turns, in principle, allows one to reach a high finite speed of the magnetic dipoles. When the electric current is flowing in the coils, there is the magnetic field which pulls the magnetic dipole inside the coil with the current. After the magnetic dipole passing through the center of the coil, the magnetic field gradient changes its sign - due to this the magnetic dipole begins to accelerate in the opposite direction, i. e. inhibits. Therefore, to create a continuous acceleration, the current in the loop must quickly break off after the magnetic dipole passing through the coil center.

However, despite the apparent simplicity of this method, its practical use is accompanied with serious difficulties.

From the ferromagnetic materials being used for magnetic dipoles the most suitable is iron having a high specific magnetic moment and high Curie temperature. The specific magnetic moment is the property of the substance and cannot be increased. Furthermore, since the magnetic dipole must include a jet engine with the fuel supply and navigation devices, the specific magnetic dipole moment will be even less than that of the pure iron. That is why it is not possible to achieve a large finite speed of the magnetic dipole by using the acceleration method.

We explain the details. The magnetic moment per molecule in iron [2], page 524, is of the $n_b = 2.219$ Bohr magneton. The table value of the Bohr magneton power [2], page 31, $\mathbf{m_b} = 9.27 * 10^{-21}$ erg / Gs. Taking into account that the atomic weight of iron is: $A_{Fe} = 56$, we find that the magnetic moment per nucleon in iron is: $\mathbf{m}_{Fe} \approx 2 * 10^{-10}$ eV / (Gs*nucleon) and this value $\mathbf{m}_{Fe}$ cannot be increased. Another principle disadvantage of this device is that the movement of the magnetic dipoles therein is unstable in the longitudinal direction. The reason is that at the attraction of the magnet poles with opposite signs there is phase instability [3].

Thus, if the magnetic dipole is slightly behind the running current pulse accelerating it, then it will turn out to be in a lower momentum field and, finally, will be forever behind it. If the magnetic dipole becomes too close to the current pulse, it will fall into a stronger field, being more and more attracted to the current pulse at the end. Finally, it will be ahead of it and turn by 180 degrees.

From the point of view of mutual positioning of the accelerating running pulse and the magnetic dipole, there is the only stable case when the pulse pushes (not pulls) the magnetic dipole. This means that the region of the phase (longitudinal) stability is located on the front slope of the traveling pulse. In the accelerator technology it is called the principle of phase stability [3].

The specific magnetic dipole moment can be, probably, increased (as compared with iron) by applying a current of the superconducting layer located inside the dipole. The magnetic field gradient, which accelerates the dipole, can be increased by superconductivity. The both ways lead to increasing the force accelerating the dipole: $F_z = \mathbf{m} * dB_z / dz$, where m - specific magnetic dipole moment, dB_z / dz - the magnetic field gradient.

We assume that the initial speed of the dipole is: $V_{sh} = 1$ km / s and it is achieved by the gas-dynamic acceleration.

1. Opportunity of increasing the specific magnetic moment in the magnetic dipole

The specific magnetic dipole moment can be increased (as compared with iron), if to place the Nb_3Sn superconducting winding inside the dipole and let the ring current flow through it.

Let us calculate how much the specific magnetic moment – the magnetic moment per unit of the mass of the magnetic dipole, will grow if to put a layer of superconducting Nb3Sn with radius $r_{cyl} = 3$ cm and thickness $\delta_{cyl} = 0.2$ cm, in its cylindrical part with a length equal to $l_{cyl} = 40$ cm. We assume the current density in the superconductor to be equal [2], page 312, to $J_{ss} = 3 * 10^5$ A/cm^2. Then, linear density j_{ss} current (A / cm) in such a superconducting layer is equal to: j_{ss} (A / cm) = $J_{sp} * \delta_{cyl} = 6 * 10^4$ A / cm. Such linear current density on the surface of the superconductor will create the magnetic field strength equal to H_{ss} (kGs) = 1.226 * j (A / cm) $\approx$ 70 kGs, that does not contradict the opportunity of achieving the current density of $J_{ss} \approx 3 * 10^5$ A/cm^2 [2], page 312. The total current flowing in the superconducting layer is equal to $I_{SC} = j_{ss} * l_{cyl} = 2.4 * 10^6$ A, it will lead to the appearance of magnetic moment $\mathbf{M}_{ss} = I_{SC} * \pi r_{cyl}^2 = 6.8 * 10^7$ A * cm^2 or, in the CGS system, $\mathbf{M}_{ss} = 6.8 * 10^6$ erg / Gs.

The total mass of the superconducting layer can be calculated from the following: the density of Nb$_3$Sn superconductor is $\rho_{Nb3Sn} = 8$ g/cm^3, the atomic mass: A = 400 and the total volume of superconductor $V_{sp} = 150$ cm^3 contains $N_{Nb3Sn} = 7.2 * 10^{26}$ nucleons. The specific magnetic moment, the magnetic moment per unit of the mass (nucleon), is then equal to: $\mathbf{M}ss = m_{ss} /N_{Nb3Sn} =$ =0.94 * 10^{-20} erg / (Gs *nucleon) = 5.9 * 10^{-9} eV / (Gs * nucleon), that is approximately 30 times greater than in iron [2], page 524.

2. Ways to achieve the required parameters of acceleration

Let the mass of the superconductor in the magnetic dipole be $m_{Nb3Sn} = 1.2$ kg, the mass of the jet engine, fuel, navigation control devices is equal $m_{Fuel} = 0.8$ kg, then the specific magnetic moment in the magnetic dipole will be equal to: $\mathbf{m_0} = 3.5 * 10^{-9}$ eV / (Gs *nucleon), that is approximately 17 times greater than in iron. The pulse duration of the magnetic field can be determined from the following considerations. To place the magnetic dipole on the length of the pulse accelerating it, slowdown wavelength must be of the order of: $\lambda_{slow} = 4$ m. The time period T_0 of the corresponding wave is determined from the following relationship:

$$\lambda_{slow} = V_{sh} *T_0, \qquad (1)$$

where we find that $T_0 = 4$ ms and the wave frequency corresponding to this period is equal to: $f_0 = 250$ Hz.

3. Selecting the thickness of the barrel wall of the Gauss gun

The region of the phase stability in azimuthally symmetric wave corresponds to the region of the radial instability. The magnetic dipole will push itself from the pole of the same sign, but, first of all, it will seek to turn around by 180^0 and be attracted by the opposite sign poles. To prevent the radial escape of the dipole and its reversal by 180^0 in the field of the pulse accelerating is possible if to place the dipole inside the titanium tube whose inner diameter is the same as the outer diameter of the dipole. The titanium tube wall thickness must be of such value to let the external magnetic field freely without distortion penetrate inside. It means that it should be much smaller than the skin-layer depth in titanium.

Electrical resistance of copper $\rho_{Cu} = 1.67 *10^{-6}$ Ohm * cm, titanium $\rho_{Ti} = 55 *10^{-6}$ Ohm *cm, [2], page 305, the conductivity σ (dimension σ is 1 / s) are related with a specific resistance value: $\sigma = 9 * 10^{11} / \rho$ for copper the conductivity is : $\sigma_{Cu} = 5.4 * 10^{17}$ 1 / s, for titanium $\sigma_{Ti} = 3.23 * 10^{16}$ 1 / s. This allows one to calculate the depth of the skin-layer and, thereby, to calculate the possible thickness of the tube wall, where the magnetic dipole will be accelerated.

Let us find the thickness of the skin - layer for titanium for frequency $f_0 = 250$ Hz. It can be calculated by the following formula:

$$\delta_{Ti} = c/2\pi \, (f_0\sigma_{Ti})^{1/2} = 1.68 \text{ cm.} \qquad (2)$$

This means that the wall thickness of the titanium tube Δh_{Ti}, wherein the magnetic dipole has to be accelerated, can be chosen to be equal to: $\Delta h_{Ti} = 2$ mm.

4. Interaction of the dipole with the magnetic field gradient

For stable acceleration of magnetic dipoles it is necessary to "switch on" consequently the magnetic coils according to the dipole move. The magnetic field of the coil with a current can be written as follows:

$$B_z = I_0 * r_0^2 / [2 * (r_0^2 + z^2)^{3/2}], \qquad (3)$$

where: I_0 - current in the loop, Ampere, r_0 - the radius of the coil with a current, cm, z - the distance from the coil plane to the observation point.

In comparison of the multi section Gauss gun [1] the corresponding coil here is necessary "to be switched on" after the magnetic dipole passage through the coil center but not to join the coil switched on in advance.

4.1. Acceleration of the magnetic dipole by the current single coil

We differentiate expression (3) with respect to z and obtain a formula for the magnetic field gradient:

$$dB_z / dz = (3/2) * r_0^2 * I_0 * z / (r_0^2 + z^2)^{5/2}. \qquad (4)$$

From this formula it is clear that the gradient field is zero in the coil plane at $z = 0$.

We assume that at a distance of the order of the radius of the turn, the speed of the dipole varies slightly, i.e. it is possible to change variable z for $V_{sh}t$. The specific magnetic dipole moment increases while the dipole passing through the center of the coil according to the law:

$$\mathbf{m} = 2\mathbf{m_0} * z / l_{sh}, \qquad (5)$$

where $\mathbf{m_0} = 3.5 * 10^{-9}$ eV / (Gs * nucleon), l_{sh} - length of the dipole . The current in the coil after switching increases linearly with time, according to the law:

$$I \approx I_0 * (4t/T_0), \qquad (6)$$

where T_0 - time period of the slowdown wavelength.

The force influencing the dipole from the coil side is:

$$F_z = \mathbf{m_0} * (z / l_{sh}) * (3/2) * r_0^2 * z * I_0 * 4 (t/T_0) / (r_0^2 + z^2)^{5/2}. \qquad (7)$$

Substituting t for z / V_{sh} and integrating over z, we obtain an expression for the energy gain rate while passing one loop by the magnetic dipole:

$$\Delta W = \int F_z dz =$$
$$= \int \mathbf{m_0} * (z / l_{sh}) * (3/2) * r_0^2 * z * I_0 * 4 (z/V_{sh}T_0) /(r_0^2 + z^2)^{5/2} dz, \qquad (8)$$

or

$$\Delta W_1 = 12\mathbf{m_0} * (r_0^2/l_{sh}) * (I_0/V_{sh}T_0) \int_0^{l_{sh}/2} [z^3 / (r_0^2 + z^2)^{5/2}] \, dz. \qquad (9)$$

Integration in (9) should be performed till the distance approximately equal to half of the length of the magnetic dipole: l_{sh} = 1m. The same order should be the turn radius: r_0 = 1m. After the magnetic dipole passing a distance l_{sh} / 2, its magnetic moment does not any longer increase and the dipole magnet will be just repelled by a coil with a current.

The corresponding set of energy can then be written as follows:

$$\Delta W_2 = 6\mathbf{m_0} * r_0^2 * (I_0/V_{sh}T_0) \int_{l_{sh}/2}^{r_0} [z_2 / (r_0^2 + z^2)^{5/2}] \, dz. \qquad (10)$$

Substituting numerical values of r_0 = 1m, l_{sh} = 1m, calculating the integrals and summing ΔW_1 and ΔW_2, we find that the energy acquired during the passage of one current loop by the magnetic dipole is equal to:

$$\Delta W \approx 12\mathbf{m_0} * (r_0^2/l_{sh}) * (I_0/V_{sh}T_0) * 5.5 * 10^{-2}. \qquad (11)$$

4.2. Acceleration of the magnetic dipole by consequence of current turns

We assume that the consequence of current turns is as follows: per 1m there are 10^3 turns (10^3 / m), the current in each coil is assumed to be equal to: I_0 = 150 kA. Assuming $\mathbf{m_0}$ = 3.5 * 10^{-9} eV / (Gs *nucleon) and averaging the action of the turns on the magnetic dipole with a coefficient of ½, we finally obtain the formula for the energy gain rate of the magnetic dipole:

$$\Delta W = 4.33 * 10^{-4} \text{ (eV / nucleon * m)}. \qquad (12)$$

Multiplying ΔW = 4.33 * 10-4 (eV / nucleon * m) by the length of the acceleration L_{acc} = 300 m, we find the finite energy of the magnetic dipoles: W_{fin} = 0.13 eV / nucleon, that corresponds to the finite speed of the magnetic dipoles V_{fin} = 5 km / s.

Figure 1 shows a diagram of the device.

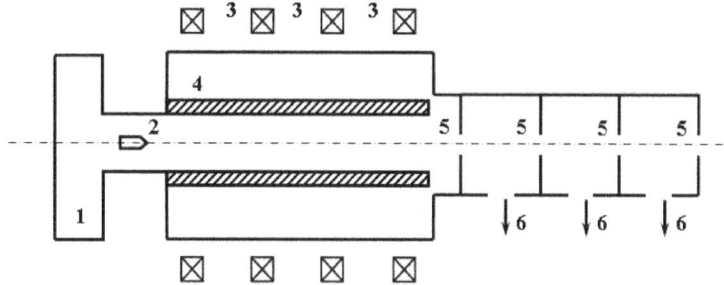

Fig.1. (1) - gun, (2) - magnetic dipoles, (3) - current coils,
(4) - titanium tube, (5) - pulse diaphragm, (6) - pumping.

Conclusion

While increasing the diameter of the magnetic dipole its total magnetic moment grows as the area of its turn, i.e. proportionally, as r^2. The dipole mass, at a constant current density increases as the perimeter of the loop, i.e., linearly with the radius. Thus, the specific magnetic dipole moment will grow linearly with increasing of the turn radius.

Literature

1.http://ru.wikipedia.org/wiki/Пушка_Гаусса

2. Tables of physical quantities. Reference ed. Kikoin,
   Moscow, Atomizdat, 1976

3. V. I. Veksler, Dokl. USSR Academy of Sciences, 1944, v. 43,
   Issue 8, p. 346, E. M. McMillan, Phys. Rev. 1945, v. 68, p. 143

Electromagnetic way of accelerating the magnetic dipoles

The article considers an opportunity of electrodynamics accelerating the magnetic dipoles at initial velocity V_{in} = 0.6 km / s, which is the magnetic dipole gain after pre-gas-dynamic acceleration to finite velocity V_{fin} = 8.5 km / s. The acceleration length L_{acc} = 2.27 km. The dipoles are accelerated at the forefront of the current pulse running through the spiral up. The accelerated magnetic dipoles have mass m = 1 kg, diameter d_{sh} = 20 mm, overall length l_{tot} = 65 cm, the length of the conical part of the l_{cone} = 20 cm. When selecting the drag coefficient C_x and the lift coefficient C_y equal to ~ 10^{-2}, the dipoles rise to height H = 10 km during a period of time τ = 14 s, thus reaching the vertical velocity V_r = 1 km / s and reducing the forward velocity till V_φ = 7.5 km / s. The magnetic dipoles reach flight range S_{max} = 12300 km.

Introduction

There is a known [1] method of accelerating the magnetic dipole - the Gauss gun - a version of the electromagnetic mass accelerator, named after the German scientist Carl Gauss, who laid the foundations of the mathematical theory of electromagnetism. The Gauss gun consists of a solenoid with a trunk inside (usually dielectric). One end of the stem is inserted into the shell (made of ferromagnetic material). When the electric current flows in the solenoid the magnetic field which appears there, accelerates the projectile, "pulling" it into the solenoid. To keep the solenoid shell against being involved into the opposite direction, i.e., not to slow down, the solenoid must be turned off at this moment. That is why for the greatest effect the current pulse in the solenoid must be short-lived and powerful. As a rule, electrical capacitors with high-voltage are used to obtain this pulse.

However, despite the apparent simplicity of the Gauss gun and obvious advantages, its practical use is accompanied with serious complications.

Among ferromagnetic materials being used in the magnetic dipoles, iron is the most suitable one because of its high specific magnetic moment and high Curie temperature. The specific magnetic moment is a property of the useable substances and can not be increased. Moreover, due to the fact that the magnetic dipole must also include the jet engine with the fuel and navigation devices, the specific magnetic dipole moment will be even less than that of the pure iron. It does not allow the magnetic dipole in the Gauss gun to reach the high finite velocity.

To increase the finite velocity of the magnetic dipole, it is possible by two ways: to increase the amplitude of the accelerating current pulse or – the specific magnetic moment. A significant increase of the amplitude of the pulse causes a strong need to increase the dielectric durability of the insulation. The superconducting coil located inside the magnetic dipole increases the specific magnetic moment by several times in comparison with a pure iron dipole. This can significantly increase the acceleration rate and finite velocity of the magnetic dipoles.

To achieve the speed of 8.5 km /s, it is necessary to increase the length of the acceleration by a few kilometers, so that the accelerator will have to be placed horizontally. The corresponding angle between the direction of the velocity and the horizon, which is needed to go up through the atmosphere, must be installed by the lift force acting on the magnetic dipole. This means that the head part of the magnetic dipole must have the corresponding asymmetry, and the magnetic dipole should have the parts, stabilizing its orientation in space.

Accelerator

1. Pre-acceleration of the magnetic dipoles with the gas-dynamic method

To speed up the magnetic dipoles by the field of the running wave, the wave must be very slow. It must be mentioned that the relative velocity $\beta = 10^{-6}$, where $\beta = V / c$, $c = 3 * 10^5$ km / s - the speed of light in vacuum, consistent with the normal speed equal to: $V = 0.3$ km / s, and it is less than the speed of the sound in the air. The gas-dynamic acceleration method does not allow one to reach the speed significantly higher than the speed of the sound in the air. For example, specifications of the gun AP 35/1000 produced by the German company "Rheinmetall" are as follows: the initial rate of shooting $V_{in} = 1.5$ km / s, the diameter of the projectile: $d_{sh} = 35$ mm. The company "Mauser" is developing an aircraft gun with a caliber (diameter of the projectile) $d_{sh} = 30 - 35$ mm and the initial projectile velocity $V_{in} = 1.8$ km / s.

To achieve a low drag coefficient of the magnetic dipole, it is required to have a form of the dipole to be a cylinder with a pointed cone at the head part. At the same time, due to a small diameter of the magnetic dipole and its great length, it will be difficult to achieve the initial velocity V_{in} like in the aircraft guns, therefore, in this case, the initial velocity of the magnetic dipole should be smaller than $V_{in} = 1.5$ km / s.

Probably, the use of the sabot projectile would allow one to achieve a greater initial velocity than we suppose.

2. Selection of the basic parameters

Let us choose the parameters of the accelerated dipole: d_{sh} = 20 mm, diameter of the total length of the dipole l_{tot} = 65 cm, the length of the conical part l_{cone} = 20 cm. The magnetic dipole is made of iron, the initial velocity of the magnetic dipole V_{in} = 0.6 km / s, the finite velocity of the magnetic dipole V_{fin} = 8.5 km / s, the gradient of the magnetic field in the pulse accelerating the magnetic dipole is taken as $\partial H_{zw} / \partial z = G$ = 2 kGs / cm.

The magnetic moment per atom in iron, [2], page 524, is of the value m_{Fe} = 2.219 Bohr magneton. The table value of the Bohr magneton is: [2], page 31, $m_b = 9.27 * 10^{-21}$ erg / Gs. Taking into account that the atomic weight of iron A_{Fe} is A_{Fe} = 56, we find that the magnetic moment per nucleon of the iron is: $m_{Fe} = 2 * 10^{-10}$ eV / (Gs * nucleon). The specific magnetic dipole moment may be increased if to place inside the dipole a superconducting coil of Nb_3Sn and let the ring current go via it.

3. The opportunity of increasing the specific magnetic moment in a magnetic dipole

We calculate how much the specific magnetic moment - the magnetic moment per unit of the magnetic dipole mass – will grow, if in its cylindrical part with a length equal to l_{cyl} = 40 cm, there is a superconducting layer of Nb_3Sn with radius r_{cyl} = 1 cm and thickness δ_{cyl} = 0.2 cm . We assume that the current density in the superconductor is equal to, [2], page 312, $j_{sc} = 3 \times 10^5$ A/cm². Then linear density J_{sc} of current (A / cm) in the superconducting layer will be J_{sc} (A / cm) $= j_{sc} * \delta_{cyl} = 6 * 10^4$ A / cm. This linear current density on the surface of the superconductor will create the magnetic field H_{sc} (kGs) = 1.226 * J (A / cm) ≈ 70 kGs, which does not contradict the opportunity of achieving the current density $j_{sc} = 3 * 10^5$ A/cm², [2], page 312.

The total current flowing in the superconducting layer is equal to: $I_{sc} = J_{sc} * l_{cyl} = 2.4 * 10^6$ A, and will cause the magnetic moment equal to $M_{sc} = I_{sc} * \pi r_{cyl}^2 = 7 * 10^6$ A * cm² or in the system CGS $M_{sc}= 7 * 10^5$ erg / Gs.

The total mass of the superconducting layer can be calculated from the fact that the density of Nb_3Sn superconductor is ρ_{Nb3Sn} = 8 g/cm³, atomic mass

42

A = 400, and the total superconductor V_{sc} = 50 cm³ contains N_{Nb3Sn} = 2.4 * 10^{26} nucleons. The specific magnetic moment, the magnetic moment per mass unit (nucleon), will be equal to the following:
m_{sc} = M_{sc} / N_{Nb3Sn} = 2 * 10^{-9} eV / (Gs * nucleon), which is about 10 times higher than in iron, [2], page 524.

Let the mass of iron in the magnetic dipole be m_{Fe} = 0.4 kg, the mass of the superconductor is also m_{Nb3Sn} = 0.4 kg, the mass of the jet engine, fuel, instruments, navigation and control devices is m_{fuel} = 0.2 kg.
Then the specific magnetic moment in the magnetic dipole will be as follows:
m_{md} = 8.8 * 10^{-10} eV / (Gs * nucleon), which is approximately by 4.4 times greater than in iron.

Let us take the gradient of the magnetic field accelerating the magnetic dipoles to be equal to: ∂H_{zw} / ∂z = 2 kGs / cm. In this case, the rate of the energy gain by the magnetic dipole will be:
ΔW_{sh} = m * ∂H_{zw} / ∂z = 1.76 * 10^{-6} eV / (cm * nucleon). To achieve the energy gain from the initial to the finite energy W_{fin}, corresponding to the finite velocity, V_{fin} = 8.5 km / s, W_{fin} = 0.4 eV / nucleon, the acceleration length L_{acc} = W_{fin} / ΔW_{sh} = 2.27 km is required.

4. Ways to achieve the desired parameters of the accelerator

Now we come to defining the spiral where the acceleration of magnetic dipoles is expected to take place with a specific magnetic moment:
m = 8.8 * 10^{-10} eV / Gs * nucleon from the initial velocity: β_{in} = 2 * 10^{-6} to the finite velocity: β_{fin} = 2.83 * 10^{-5}, β = V / c, c = 3 x 10^{10} cm / s - velocity of light in vacuum.

The radii of the spiral, [3], the initial and finite ones are taken as follows: r_{0in} = 50 cm, r_{0fin} = 30 cm. The accelerator, due to damping, will be divided into sections. Therefore, within one section it is possible to choose the initial and finite values of the parameter x = $2\pi r_0/\beta\lambda_0$, close to the optimal ones and equal to x = 6.28 * 50 / (2 * 10^{-6} * 1.3 * 10^8) = 1.2.

Selection of wavelength λ_0 = 1.3 * 10^8 means that we have chosen the pulse duration to be equal to the following:
(f_0 = c/λ_0 = 230 Hz), τ_{pulse} = 1 / (2f_0) = 2.17 ms.

The slowdown wavelength is equal to the following:

$\beta\lambda_0 = 2 * 10^{-6} * 1.3 * 10^8 = 260$ cm, and the magnetic field gradient $\partial H_{zw} / \partial z = G = 2$ kGs / cm, and corresponds to the amplitude of the magnetic field pulse: $H_{zw} = 82.8$ kGs. This amplitude of the magnetic field can be found from the relation $\partial H_{zw} / \partial z = k_3 H_{zw} = 2\pi H_{zw}/\beta\lambda_0$. From this we get the value $H_{zw} = \beta\lambda_0 * G/2\pi = 82.8$ kGs.

To find the power flux needed to generate the magnetic field of this strength, we find a relation between the components of the electric field $E_{zw} = E_0 I_0 (k_1 r)$ and of the magnetic field:
$H_{zw} = (k_1 / k) \, tg\Psi I_0 (k_1 r_0) E_0 I_0 (k_1 r) / I_1 (k_1 r_0)$, [3]. For the interior of the spiral, where k_1 is the transverse wave vector: $k = (\omega / c) * \varepsilon^{1/2}$ - the wave vector, r_0 - the radius of the spiral, the expression is as follows: $(k_1 / k) = 1/\beta_{ph}$, $tg\Psi \approx h/2\pi r_0$. So, $(k_1 / k) * tg\Psi = \varepsilon^{1/2}$ at the beginning of the spiral $k_1 r_0 = 1.2$ and the ratio of $I_0 (k_1 r_0) / I_1 (k_1 r_0) = 2$. Thus, the component of the magnetic field $H_{zw} = 82.8$ kGs on the spiral axis that corresponds to the electric field strength: $E_{zw} \approx 347.2$ kV / cm.

The value in the brackets, {}, (in formulae (13) in [3]), for the value of the argument x = 1.2,
$\{\} = \{(1 + I_0 K_1/I_1 K_0) (I_1^2 - I_0 I_2) + \varepsilon (I_0/K_0)^2 (1 + I_1 K_0 / I_0 K_1) (K_0 K_2 - K_1^2)\}$ is:
$\{\} = 3,77 * \varepsilon$ so that the power required to achieve the electric field strength $E_{zw} = 347.2$ kV / cm for the initial speed of the magnetic dipole $\beta_{in} = 2 * 10^{-6}$, may be found from formula [3]:

$$P = (c / 8) * E_{zw}^2 * r_0^2 * \beta * \{\}. \qquad (1)$$

The wave power, in Watts, is equal to [3]:

$$P (W) = 3 * 10^{10} * (3.47)^2 * 10^{10} * 2.5 * 10^3 * 2 * 10^{-6} * 1.28*10^3 * 3.77 /$$
$$(8 * 9 * 10^4 * 10^7) = 12.23 \text{ GW}. \qquad (2)$$

According to formula (2), to achieve the magnetic field gradient on the axis of the field G = 2 kGs / cm, it is required to have power P = 12.23 GW. This power can be achieved by using the pulse technology.

We introduce the notion of pulse amplitude $\tilde{U}_{acc}$ related with the field power on the axis of the spiral E_{0pulse} by the following ratios [3]:

$$\tilde{U}_{acc} = E_{0pulse}\lambda_{slow}/2\pi, \quad \lambda_{slow} = \beta\lambda_0, \quad \lambda_0 = c/f_0. \qquad (3)$$

44

Thus, the amplitude of the power pulse propagating along the spiral must be equal: $\tilde{U}_{acc} = E_{0pulse} * \lambda_{slow}/2\pi = 14.37$ MV.

Table 1 summarizes the main parameters of the accelerator of magnetic dipoles.

Table1. Parameters of the accelerator.

Parameter	Value
m = $8.8*10^{-10}$, the dielectric outside spiral, ∂ Hz / ∂ z = 2 kGs / cm, wave power, P	P = 12.23 GW, $\mu = 1$, $\varepsilon = 1280$
Velocity, initial –finite, β_{ph}	$\beta_{ph} = 2*10^{-6} - 2.83*10^{-5}$
The radius of the spiral, initial -finite, r_0	$r_0 = 50 - 30$ cm
Frequency of the wave, f_0	$f_0 = 230$ Hz
The electrical field strength, E_{zw}	$E_{zw} = 347$ kV/cm
Accelerator length, L_{acc}	$L_{acc} = 2.27$ km
Pulse duration, τ	$\tau = 2.17$ ms
The amplitude of the voltage $\tilde{U}_{acc}$	$\tilde{U}_{acc} = 14.37$ MV

5. Preventing the turn of the dipole by 180⁰ in the magnetic field pulse by imposing the uniform magnetic field

Everything would have been well if the dipole were a point. But the magnetic dipole is not a point and the action on it by the radial component of the magnetic field leads to "roll over", a reversal position of the dipole. Two interacting coils will seek to be placed in such a way that their planes should be parallel to each other, and the direction of the currents of the both would be the same. In contrast to the forces accelerating the dipole and leading to the radial displacement of the center of mass of the dipole, the action of the pair of the forces, leading to the "reversal" of the dipole, is summed up.

The simplest solution that prevents the turn of the dipole by 180⁰ in the accelerating field of the wave is the imposition of the uniform external magnetic field which won't influence on the acceleration of the dipole because of its homogeneity but will only hold the dipole against the reversal in the space.

Impose the external magnetic field H_{out} on the spiral with the dipole to keep it against turning off. The magnetic field of the wave is of the order of $H_{zw} = 82.8$ kGs, respectively, to compensate small deviations $\sin\theta <1$, the external magnetic field must be $H_{out} > 100$ kGs.

Thus, we have fixed the magnetic dipole orientation in the space. Now the reversal moment will act on the turns of the coils, which carry the impulse current, and therefore they must be reliably fixed. One can consider a mechanical model of such an accelerator, which should consist of two magnets directed to each other with the same sign poles. One of the magnets - is a magnetized iron magnetic dipole. The second magnet - is the current pulse running via the coils of the spiral. The one-sign poles of the magnets repel each other - the current pulse in the spiral is speeding up and accelerates (pushes) the magnetic dipole.

In this accelerator technology it is called the automatic phase stability principle [4]. Thus, the running current pulse can only push (not pull) the dipole being accelerated.

A dedicated narrow channel (trunk) holds the magnet being pushed against turning over. In our case the constant external magnetic field fulfills the role of this channel. The pushing magnet cannot "turn around" either; it is prevented from turning by mechanical forces.

6. Radial focusing by the magnetic field having a variable component

To eliminate the opportunity for the magnetic dipoles to "escape" along the radius, we have to introduce the radial focusing. The easiest way to do it is to impose an additional sinusoidal field over the external uniform field H_{out} , so that the total field will be a combination of the constant field H_{out0} holding the magnetic dipole from the "turn around", and of the variable component of $H_{out1} * \sin k_z z$, which automatically appears when the solenoid coils producing the longitudinal magnetic field H_{out}, are positioned quite seldom. Thus, $k_z = 2\pi/a_z$, where a_z is a spatial period of positioning the coils. This ripple, corrugated $H_{out} = H_{out0} + H_{out1} * \sin k_z z$ of the magnetic field leads to the appearance of sign-variable gradient of the external magnetic field and the appearance of sign-alternative forces in the equation of radial motion, assuming $k \perp \approx k_z$,

$$r_1{''}= \omega_r^2 r_1 + \tfrac{1}{2} W_{\lambda m}\sin k_z V_z t * [H_{out1} / (H_{out0} - H_{zw})] * k_z r_{b0} * (k_z c)^2 r_1, \quad (4)$$

where $k_z V_z$ - oscillation frequency in the external magnetic field.

We change variables $k_z V_z t = \omega_{out} t = 2\tau$, then $\partial / \partial t = \tfrac{1}{2} \omega_{out} \partial / \partial \tau$, and the equation (4) can be written as follows:

$$r_1 \text{''} = \{4\,(\omega_r^2/\omega_{out}^2) -$$

$$-2W_{\lambda m} * ([H_{out1} / (H_{out0} - H_{zw})] * k_z r_{b0} * (k_z c)^2 / * \omega_{out}^2) * \sin 2\tau\}\,r_1, \quad (5)$$

and, thus, the equation is reduced to the Mathieu equation, [5], which describes the transverse motion:

$$r\text{''} + (a - 2q * \sin 2\tau)\,r = 0, \quad (6)$$

where

$$a = 4\,(\omega_r / \omega_{out})^2, \quad (7)$$

$$q = W_{\lambda m} * [H_{out1} / (H_{out0} - H_{zw})] * k_z r_{b0} * (k_z c)^2 / \omega_{out}^2. \quad (8)$$

Depending on the values of parameters a and q, in the Mathieu equation there are areas of stable and unstable solutions. Choosing the alignment of the coils forming the magnetic field (k_z) and the current in them (H_{out1}) in such a way that the solution of (6) will be within the range of sustainability and it will be possible to obtain a stable acceleration of the magnetic dipoles. The stability condition is as follows [5]: 0.92> q> a. simplification by $W_{\lambda m}$, $k_z r_{b0}$, ω_{out}^2 and $(H_{out0} - H_{zw})^{-1}$, we get

$$(k_z c)^2 H_{out1} > 8(\pi f_0/\beta_z)^2 * H_{zw}, \quad (9)$$

$$H_{out1} > 8(k_3/k_z)*(\pi f_0/\beta_z k_z c)^2\,H_{zw}, \quad (10)$$

$$H_{out1} > 4\pi(a_z/\lambda_{slow})^2\,H_{zw}. \quad (11)$$

The amplitude of the alternating field must exceed the field of a wave multiplied by $4\pi\,(a_z / \lambda_{slow})^2$.

Perhaps, small focusing magnetic elements can be placed close to the axis of the spiral waveguide. Just as in the drift tube linear accelerator of Alvarez, the magnetic elements, maybe, won't greatly affect the distribution of the accelerating pulse propagating via the spiral waveguide. Then the spatial period of the magnetic field changes can be made small enough, $a_z \approx 10$ cm.

Substituting the previous expression with the values: $H_{zw} = 82.8$ kGs,

47

$f_0 = 230$ Hz, $\beta_z = 2 * 10^{-6}$, $a_z \approx 10$ cm, $c = 3 \times 10^{10}$ cm / s - velocity of light in vacuum, we obtain:

$$H_{out1} \geq 1.85 * 10^{-2} * H_{zw} = 1.5 \text{ kGs.} \qquad (12)$$

This value of the variable component of the magnetic field will generate the sign-variable field on the basis of permanent magnets of NdFeB, similar to how it is done to focus the electron beams in traveling-wave tubes.

You can even eliminate the static magnetic field, but, nevertheless, hold the dipole against the reversal over the radius to carry out this only by the sign-variable magnetic field that has no DC component. This result is obtained if the amplitude of the external field H_{out} is 2 times larger than the pulse field H_{zw}, multiplied by the ratio of half-cycle of the external field a_z to the spatial length of the pulse l_p: $H_{out} > 2 (a_z / l_p) H_{zw}$.

7. The power damping of the pulse propagating in the spiral

Wave attenuation in a spiral waveguide will lead to the fact that the amplitude of the pulse traveling along the spiral will decrease while pulse moving from the beginning to the end of the spiral and this power reduction is related with the ohm currents for heating the spiral.

Current I_φ flows through the windings of the spiral and, in fact, it is Ohm losses:

$$\Delta P \text{ (W / turn)} = \frac{1}{2} I_\varphi^2 * R, \qquad (13)$$

where I_φ - the current going via the coil in amperes, R - loop resistance in ohms. Then ΔR / turn - is expressed in Watts.

We first find the resistance of the coil. The resistance is calculated by the usual formula: $R = \rho l / s$, where $\rho = 1.7 * 10^{-6}$ Ohm * cm - the resistivity of copper, the copper coil will be considered, $l = 2\pi r_0$ - coil length, r_0 - the radius of the spiral, s – the coil cross-section. Since the current flowing via the coil is of high-frequency (AC), a factor, ½ appears there and this current penetrates into the conductor for the depth of the skin – layer which has to be found.

The expression for the depth of the skin - the layer can be written as follows:

$$\delta = a / (\sqrt{2\pi\sigma\omega_0}), \qquad (14)$$

where $c = 3 \times 10^{10}$ cm / s - the speed of light in vacuum, $\sigma = 5.4 * 10^{17}$ 1 / s - the conductivity of copper, $\omega_0 = 2\pi f_0$ - circular frequency, $f_0 = 260$ Hz - frequency of the wave propagating in the spiral. Substituting numerical values in the formula (14) gives $\delta = 0.4$ cm.

The result is the depth of the skin - layer $\delta = 0.4$ cm, that is much larger than the distance between the turns of the spiral h = 1 / n = 0.02 cm, n $\approx$ 50 - the number of turns of the spiral per 1 cm length of the spiral. This means that in order to reduce the resistance of one turn, and, accordingly, - reduce the attenuation, it is necessary to wind the coil as a rather wide tape with a width H = 2δ = 0.8 cm. The tape should be placed on the wide side H over the radius. The distance h between the coils consists of h/2 turns and h/2 space between the coils, so that the winding pitch of the value h / 2 revolution, and the turn occupied the space h / 2, to be equal to the space between the coils.

Then the resistance of one coil $R = \rho l / s$ will be:

$$R = \rho * 2\pi r_0 / (2\delta * h / 2) = \rho * 2\pi r_0 / (\delta * h). \qquad (15)$$

Substituting numerical values for the start of spiral $r_0 = 50$ cm, we find

$$R = 1.7 * 10^{-6} * 6.28 * 50 / (2 * 0.4 * 10^{-2}) = 6.6 * 10^{-2} \text{ Ohm.} \qquad (16)$$

Now we have to find I_φ - current in the coils. To do this, we use the formula

$$H_{zsurf} = (4\pi / c)\, nI_\varphi, \qquad (17)$$

where H_{zsurf}- the magnetic field on the surface of the spiral, $H_{zsurf} \approx H_{zw}$. Now we can find current nI_φ flowing via the coils of the spiral:
nI_φ (A / cm) = H_{zsurf} / (4π / c) = $(1.226)^{-1} * H_{zsurf}$(A / cm) = H_{zsurf} (Gs).
Then – the current in one coil is:

$$I_\varphi \text{ (A)} = H_{zsurf} \text{ (Gs) / n.} \qquad (18)$$

Substituting numerical values into the formula (18) we have found that the current in one coil is:
I_φ (A) = [82.8 kA / cm] / (50 turns / cm) = 1.656 kA / revolution.

Ohm losses of the current in one coil are as follows:

$$\Delta P \ (W / turn) = \tfrac{1}{2} \ I_\varphi^2 * R = 90 \ kW / coil. \tag{19}$$

Since there are n turns per 1 cm, the energy losses per 1cm will be by n times more:

$$\Delta P \ (W / cm) = \tfrac{1}{2} \ I_\varphi^2 * R * n = 4.5 * 10^6 \ W / cm. \tag{20}$$

We introduce the ratio

$$\Delta P / P = -2\alpha, \tag{21}$$

hence,

$$1 / \alpha = L_{damping} = 2P/\Delta P = 2 * 12.23 * 10^9/4.5 * 10^6 =$$
$$= 5.43 * 10^3 \ cm = 54 \ m, \tag{22}$$

is the length at which the field power is reduced by a factor of e due to damping. It can be seen that the movement of the magnetic dipole while accelerating is necessary to calculate taking into account the damping of the power pulse propagating via the spiral waveguide.

8. Magnetic dipole capture in the acceleration mode. Excesses

We calculate the required accuracy of coincidence of the initial phase of the accelerating wave (pulse) with the synchronous phase. The theory of particle capture in a traveling wave gives, [6]: $\Delta\varphi = 3\varphi_s$, $(+ \varphi_s - 2\varphi_s)$. In practice, it means, for example, in our case, where $\lambda / 4$ corresponds to the duration of 1 ms or 90^0, that one degree corresponds to the time interval of approximately 10 μs. In linear accelerators the buncher gives the phase width of the bunch $\pm 15^0$, and to avoid large phase fluctuations, we require that the synchronization accuracy of the magnetic dipole with the accelerating pulse was as follows: $\Delta\tau = \pm 15 * 10 \ μs = \pm 150 \ μs$. Such synchronization accuracy seems to be quite reachable for the gunpowder start, i.e., - preliminary gas-dynamic acceleration of the magnetic dipoles.

Let us now calculate the excess for the accuracy of coincidences of the initial velocity of the magnetic dipole and the phase velocity of the pulse propagating along the spiral structure. We introduce value $g = (p-p_s) / p_s$ - the relative difference between the pulses [6]. In the non-relativistic case – it is just a relative velocity spread out of $g = (V-V_s) / V_s$. The vertical range of the separatrix is calculated with the following formula [6]:

$$g_{max} = \pm 2 (\Omega_s/\omega_0) [1 - \varphi_s / ctg\varphi_s]^{1/2}, \qquad (23)$$

wherein: $\varphi_s = 45^0 = \pi / 4$, $ctg\varphi_s = 1$, $[1 - \varphi_s / ctg\varphi_s]^{1/2} = 0.46$, $2 * 0.46 = 0.9$ $\omega_0 = 2\pi f_0 = 1.44 * 10^3$, $\Omega_s/\omega_0 = [W_{\lambda m} ctg\varphi_s / 2\pi\beta_s]^{1/2}$.

Let us determine the value of $W_{\lambda m} = (m \, \partial \, Hz \, / \, \partial \, z) \, \lambda_0 sin\varphi_s / Mc^2$ - a set of the relative magnetic dipole energy at the wavelength λ_0 in vacuum. In our case $\lambda_0 = c/f_0 = 1.3 * 10^8$ cm, $sin\varphi_s = 0.7$, $Mc^2 = 1$ GeV, $W_{\lambda m} = 3.64 * 10^{-8}$, $ctg\varphi_s = 1$. Substituting numerical values we get $g = (V_{in} - V_s) / V_s = \Delta V / V_s$, and, finally, $\Delta V / V_s = \pm [3.64 * 10^{-8} / (6.28 * 10^{-6} * 2)]^{1/2} * 0.9 = \pm 0.05$.

Thus, the allowed discrepancy of the initial speed of the magnetic dipole with the pulse transfer is of the order of $\Delta V / V_s = \pm 5\%$.

9. Magnetic dipoles from vacuum to atmosphere

Magnetic dipoles should be accelerated in a rather high vacuum $P_1 \approx 10^{-6}$ mm. Hg. while their use is expected under normal atmospheric conditions: $P_2 \approx 10^3$ mm. Hg. So, the pressure difference is about nine orders of the magnitude. To reach such a pressure gradient, we can use several buffer cavities which are cylindrical chambers separated from each other by walls with the imbedded pulse iris diaphragms. Each cavity should be supplied with individual vacuum pumping.

We calculate the amount of the air particles which penetrate to the first cavity - the nearest one to the atmosphere. Let the radius of the iris be $r_{0d} = 10$ cm, and the time to open it $t_1 = 10^{-3}$ s. Then the linear velocity of the iris petals will be $V_{11} = r_{0d} / t_1 = 10^4$ cm / s, which should not cause a problem for the operation of the mechanism. The average velocity of the thermal motion of the molecules of air V is assumed roughly to be equal to the speed of sound in the air: $V = 3 * 10^4$ cm / s. From the various spatial orientations of the velocity only 1/6 (1/6 - one facet of the cube) is directed towards the aperture. The number of molecules per cubic centimeter of air under normal conditions, the number of Loschmidt is $\rho_{0L} = 2.7 * 10^{19}$ molecule/cm³. Then the number of molecules from the atmosphere which penetrated to the first buffer cavity when the diaphragm was open, will be as follows:

$$N_0 = (1/6) \, \rho_{0L} * \pi r_{0d}^2 * V t_1. \qquad (24)$$

Substituting the numerical values into formula (24) we find that the total number of particles coming from the atmosphere to the first buffer cavity is: $N_0 = 4 * 10^{22}$ particles. Let the volume of the first cavity is the value equal to $V_{01} = 10^3 \, l = 10^6 \, cm^3$. Then the density of molecules in it after the diaphragm shutdown is equal to $n_0 = 4 * 10^{16}$ molecule/cm^3.

Particle density (and pressure: $p = nkT$) in the first buffer cavity is about 3 orders lower than the density of particles in the atmosphere under normal conditions. So, at least, three cavities of this volume will be required to reach the appropriate pressure gradient.

Now we consider the dynamics of the particle density in the cavity for the period of time between the shutdowns of the trigger device. Let the cavity be pumped out through the hole of the square of $S_1 = 10^4 \, cm^2$. We assume that all the molecules which got into this area are removed from the volume. It is supposed that the time between the cycles is $t_2 = 10^{-1} \, s$, it means that the frequency of operation of the device is equal to $F = 10$ Hz. The equation describing the reduction of the density of the particles while pumping out can be written as follows:

$$dn = -n * (1/6) * S_1 V t / V_{01}. \qquad (25)$$

The solution of this equation can be written as

$$n = n_0 * \exp [- * (1/6) * S_1 V t / V_{01}]. \qquad (26)$$

For evacuation time $t = t_2$ the exponent is approximately equal to 5, thus, due to pumping out the density of molecules in the first buffer cavity is reduced by more than 100-times, $n = n_0 * 7 * 10^{-3}$. The density of particles in the first volume of the buffer before the next shutdown will be the value of $n_1 = 4 * 10^{16} * 7 * 3 * 10^{-3} = 10^{14}$ molecule/cm^3, that is by 5 orders of magnitude less than the number of Loschmidt: $\rho_{0L} = 2.7 * 10^{19}$ molecule/cm^3, corresponding to the number of molecules per cubic centimeter of the air under normal conditions.

It is evident that before the next shutdown the cavity can be considered to be empty.

Figure 1 shows a possible scheme of the device.

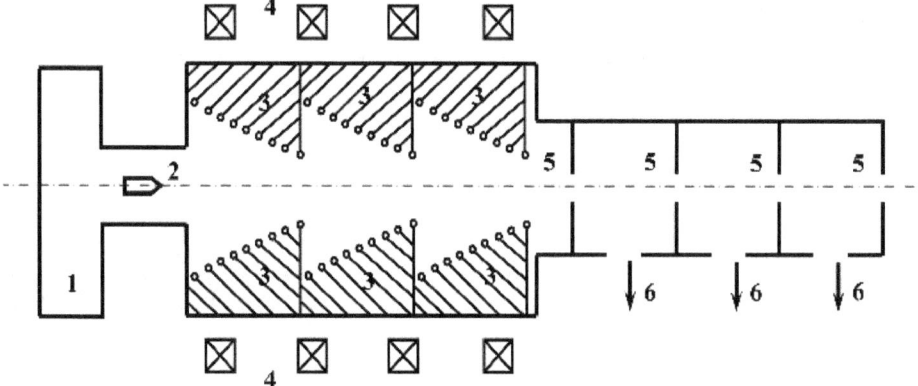

Fig.1. 1 – gun, 2 – dipoles, 3 - spiral section of the waveguide,
4 - current coils, 5 - pulse aperture, 6 - individual evacuation buffer cavities.

Applications

1. Lifting power

When the length of the accelerator $L_{acc} \approx 2.27$ km it can be positioned only horizontally. To transfer the magnetic dipole above the atmosphere, you can use a small asymmetry in the form of a magnetic dipole to create the lifting power F_y. The equation of the vertical motion in this case may be written as follows:

$$mdV_y / dt = C_y \rho_0 V_x^2 * S_{tr} / 2, \tag{27}$$

where C_y is an aerodynamic lift coefficient, $\rho_0 = 1.3 * 10^{-3}$ g/cm³ - the air density on the surface of the Earth, $V_x = 8.5$ km / s - the horizontal speed of the magnetic dipole, S_{tr} – the transversal cross-section of the magnetic dipole.

It is required that during $t_{fly} = 10$ s the magnetic dipole rises by a height of $H_{fly} = 10$ km, where the air resistance is negligible. From (27) we find the proper lifting coefficient C_y for this case:

$$C_y = 4mH_{fly} / (\rho_0 V_x^2 * S_{tr} * t_{fly}^2). \tag{28}$$

Substituting the numbers into the expression (28): $m = 10^3$ g, $S_{tr} = 3.14$ cm², we find that the lifting coefficient C_y must be equal to $C_y \approx 10^{-2}$, that is, probably, not difficult to fulfill by a bent angle in the head part of the magnetic dipole.

2. Ballistics. Air resistance

It is necessary to calculate the motion of magnetic dipoles accelerated by using the electrodynamics method. The equation of motion of magnetic dipoles can be written as

$$m dV / dt = - \rho C_x S_{tr} V^2 / 2, \tag{29}$$

where m is the mass of the magnetic dipole, V-velocity, $\rho = \rho_0 e^{-z/H0}$ - barometric formula of changing the atmospheric density with altitude, $\rho_0 = 1.3 * 10^{-3}$ g/cm^3 - the air density at the surface of the Earth, $H_0 = 7$ km - the altitude at which the density drops by factor e.

The aerodynamic coefficient or drag coefficient is called a dimensionless value that takes into account the "quality" of the form of the magnetic dipole:

$$C_x = F_x / (\tfrac{1}{2}) \rho_0 V_0^2 S_{tr}. \tag{30}$$

The solution of equation (29) can be written as follows:

$$V(t) = V_0 / [1 + \rho C_x V_0 * S_{tr} * t/2m]. \tag{31}$$

To calculate the change of the speed of magnetic dipoles, it is necessary to find the aerodynamic coefficient C_x.

3. The calculation of the drag coefficient of magnetic dipoles for air

It is assumed that the magnetic dipole has the form of a cylindrical rod with a conical head part. Then, at the hit of a nitrogen molecule on a sharp cone, the change of the longitudinal velocity of the molecules is equal to

$$\Delta V_x = V_x * \Theta_h^2 / 2, \tag{32}$$

where Θ_h is the angle at the vertex of the cone. The gas molecules transfer the momentum to the magnetic dipole:

$$p = mV = \rho V_x S_{tr} t * \Delta V_x. \tag{33}$$

The change in the momentum per unit of time is the power of the frontal inhibition-

$$F_{x1} = (½) \rho V_x S_{tr} * V_x * \Theta_h^2. \tag{34}$$

Dividing F_{x1} by $(½) \rho V^2_x S_{tr}$, we get the drag coefficient for a sharp cone in the mirror reflection of the molecules from the cone (the Newton formula):

$$C_{x\,air} = \Theta_h^2. \tag{35}$$

Let the length of the cone part of the magnetic dipole is $l_{cone} = 20$ cm at diameter $d_{sh} = 20$ mm. This means that the angle at the vertex of the cone is $\Theta_t = 10^{-1}$ and $C_{x\,air} = 10^{-2}$.

In order to have a sharp cone in the head part of the magnetic dipole, it must be long enough. Limiting the length of the magnetic dipole means that for a good efficiency of its acceleration, the length of magnetic dipole l_{tot} should be less than the slowdown wavelength quarter $\lambda_{slow} = \beta\lambda_0$, i.e.: $l_{tot} < \beta\lambda_0 / 4$. In our case, to start the acceleration, $\beta\lambda_0 / 4$ is $= 65$ cm.

4. Passage of the magnetic dipoles through the atmosphere

We set a table of the time dependence of the vertical velocity of the magnetic dipole, its lifting altitude and horizontal speed. The vertical velocity is calculated by the following formula:

$$\Delta V_y = C_y \rho V_x^2 * S_{tr} * \Delta t/2m. \tag{36}$$

The altitude taking, respectively, is calculated as

$$H_{fly\,n+1} = H_{fly\,n} + V^-_y * \Delta t + C_y \rho V_x^2 * S_{tr} * (\Delta t)^2/4m, \tag{37}$$

where V^-_y is the average vertical velocity in the time interval Δt. Reducing the horizontal velocity within the time will be described by formula (31):

$$V_{x\,n+1} = V_{x\,n} / [1 + (C_x \rho V_{xn} * S_{tr} * \Delta t/2m)]. \tag{38}$$

The change in the air density with the altitude will take into account on the barometric formula $\rho = \rho_0 * \exp[-y/H_0]$, where $H_0 = 7$ km. Table 2 shows the flight parameters of the magnetic dipole depending on time. The second column shows the vertical velocity of the magnetic dipole, in the third – there is the gain altitude, in the fourth - the horizontal speed of the magnetic dipole, which it will obtain after the corresponding second of the flight.

Table 2. Dependence of the parameters on the time of flight

t, s	V_y, km/s	H, km	V_x, km/s
0	0	0	8.5
1	0.144	0.144	8.36
2	0.284	0.428	8.22
3	0.42	0.843	8.11
4	0.55	1.4	8
5	0.65	2.05	7.9
6	0.743	2.8	7.82
7	0.8	3.6	7.76
8	0.86	4.46	7.7
9	0.91	5.37	7.65
10	0.954	6.32	7.6
11	0.992	7.3	7.56
12	1.024	8.3	7.53
13	1.052	9.37	7.5
14	1.075	10.44	7.48

5. Ballistics. Flight Range

For large values of velocity $V_0 \approx 7.5$ km / s, the Earth can not be considered flat. Let us put down the equation of motion of the magnetic dipole in the cylindrical coordinate system. In this case the vertical direction will now be radial and the horizontal one - azimuthal:

$$mdV_r / dt = - mg + mV_\varphi^2/R_E, \qquad (39)$$

where $R_E = 6400$ km - the radius of the Earth, $g = 10^{-2}$ km/s^2 - acceleration due to gravity. Equation (39) can be reduced to:

$$dV_r / dt = -g (1 - V_\varphi^2/R_E g) = -g *. \qquad (40)$$

Equation (40) is $V_r = - g * t$, and, as in the case of a stone thrown at an angle to the horizon in the vacuum in the flat case, we find that the time of raising till the maximum distance and coming back to the initial point is equal to:

$$t_{max} = 2V_r / g *. \tag{41}$$

For the azimuthal velocity $V_\varphi = (R_E g)^{1/2} = 8$ km / s, the time is infinite. This means that the magnetic dipole at this velocity equal to the first space velocity will rotate along a circular orbit, and will not fall back to the Earth.

For the parameters of the magnetic dipole: $V_r = 1$ km / s and $V_\varphi = 7.5$ km / s the time t_{max} of lifting to the maximal height and return to the starting point is: $t_{max} = 1650$ s and, therefore, the flight range of the magnetic dipole is equal to $S_{max} = V_\varphi * t_{max} = 12300$ km. If to increase the length of the accelerator and, consequently, the finite velocity of the magnetic dipoles, their velocity after passing through the atmosphere will be more than 8 km / s and, thus, the dipoles are displayed onto near-the Earth orbit.

Conclusion

The magnetic moment of a current-carrying coil increases as the square of the coil that is the square of the radius growth. The coil perimeter, its mass grows with the radius increasing linearly, so that the specific magnetic moment, the magnetic moment per nucleon in the loop will grow linearly with the radius of the magnetic dipole. To limit the growth of the radius will be necessary for the sake of the sharp cone in the head part of the magnetic dipole. To have a pointed cone with the growing radius of the dipole, it will be necessary to increase its length.

You can give up using the constant magnetic field holding the magnetic dipoles against the turn by 180^0 and alternating magnetic field keeping the magnetic dipoles near the axis of the acceleration, if the acceleration is carried out in a narrow trunk. This trunk can be made of a titanium thin walled tube with a wall thickness $\Delta h_w \approx 2$ mm. However, it is not clear how small it is necessary to choose the synchronous phase to resist the force of friction, which is not regular, between the projectile and the walls of the trunk.

Literature

1. http://ru.wikipedia.org/wiki/Пушка_Гаусса

2. Tables of physical quantities, Directory Ed. I. K. Kikoin, Moscow, Atomizdat, 1976

3. S. N. Dolya, K. A. Reshetnikova, On the electrodynamics' acceleration of macroscopic particles, Communication JINR P9-2009-110, Dubna, 2009, http://www1.jinr.ru/Preprints/2009/110(P9-2009-110).pdf http://arxiv.org/ftp/arxiv/papers/0908/0908.0795.pdf

4. V. I. Veksler, Reports of the Acad. of Sci. of the USSR, v. 43, p. 329, 1944, E. M. McMillan, Phys. Rev. v. 68, p. 143, 1945

5. H. B. Mack - Lachlan, Theory and Application of Mathieu functions, Trans. from English, Moscow, Publ. Foreign Lit., 1953.

6. I. M. Kapchinsky, Particle dynamics in linear resonance accelerators, Moscow, Atomizdat, 1966

Electrodynamics acceleration of electrical dipoles

This article considers the acceleration of electric dipoles consisting of thin metal plates and dielectric (barium titanate). The dipoles are of a cylindrical shape with a diameter of the cylinder d_{out} = 2 cm and length d_d = 1 sm. Capacity of the parallel-plate capacitor is: C = 278 pF, and it is charged up to the voltage of U = 280 kV. Pre-acceleration of the electric dipoles till velocity V_{in} = 1 km /s is reached by the gas-dynamic method. The finite acceleration is produced in a spiral waveguide, where the pulse is travelling with voltage amplitude U_{acc} = 700 kV and power P = 125 MW. This pulse travels via the spiral waveguide and accelerates the injected electric dipoles in the longitudinal direction till the finite velocity V_{fin} = 8.5 km / s over length L_{acc} = 0.77 km.

Introduction

There is a known [1] method of accelerating the magnetic dipoles which enables one to accelerate the magnetic dipoles by the running current pulse. To increase the specific magnetic moment, inside the dipoles we place a superconducting coil with the excited current in it.

However, this method of accelerating the magnetic dipoles has a serious drawback. During all the period of acceleration and time of flight to the target, it is necessary to keep the low temperature and superconductivity inside the magnetic dipole. Otherwise, due to a large energy release in the magnetic dipole, it can just collapse.

Another known [2] method of accelerating the charged bodies is as follows: the body is pre-accelerated up to the velocity corresponding to the velocity of injection into the spiral waveguide, then the body is irradiated with an electron beam injected from the electron accelerator, it is electrically charged and finally accelerated by the same voltage pulse running inside the coils of the spiral waveguide.

To set several electrons onto the body is not a problem, but further on when there are many electrons on the body, they will begin to run away from it due to the auto electron emission. Let the field strength for electron emission be as follows: E = 3 * 10^7 V / cm. After reaching this field strength, no matter how many electrons you put on the body, they will flow away from the body due to the Coulomb repulsion.

Having planted enough electrons, to plant more, it is necessary to overcome the repulsion of those which are already there. This means that the energy of the electrons, which we want to put onto the body, should be large enough so that they can overcome this Coulomb barrier, reach the body and stay on it.

Coulomb barrier grows for the particles of the cylindrical form linearly with increasing of the diameter and for particles with diameter d = 20 mm it will reach 30 MeV. To overcome it, it will be necessary to accelerate electrons in a specialized accelerator.

The essence of this proposal is to consider the acceleration of the electric dipoles but not the charged bodies.

The acceleration rate for this case has been found to be equal to the following: $\Delta W / \Delta z = (N_e / A) e * (2\pi d_d/\lambda_s) * E_{0zw} * \sin\varphi_s$, where $\Delta W / \Delta z$ - a set of plate-capacitor energy per unit of the length, eN_e / A - electrical charge of any sign, located on a plate capacitor and per nucleon in it, d_d - the distance between the plates of a parallel-plate capacitor, λ_s - slow down wavelength in the spiral waveguide, E_{0zw} - the amplitude of the wave, φ_s - synchronous phase. Acceleration of the plate capacitor is produced in the dielectric channel, which prevents the turn of the dipole by 180 degrees and its deviation from the axis of the acceleration.

1. Acceleration of electric dipoles

On the electric dipoles it is possible to set a large number of related charges, i.e., the charges having opposite signs, and, thus, we get a plate capacitor. The total electric charge in the capacitor will be equal to zero, but such electric dipoles will possess a rather large electric dipole moment which can interact with the electric field gradient.

1. 1.Parameters of the electric dipole

Let us consider the acceleration of the cylindrical capacitor having an outer diameter $d_{out} = 2$ cm, the distance between the electrodes is equal to: $d_d = 1$ cm, where as dielectric we use ceramic capacitor T-900, with a relative permittivity $\varepsilon = 10^3$ and a breakdown voltage U = 28 kV / mm [3], page 321. We assume the density of the ceramic capacitor to be equal to $\rho = 6$ g/cm^3 [4].

To find the specific electric charge of the capacitor, we neglect the mass of the metal plates. The square and volume of the dielectric are:

$S_d = \pi d_{out}^2 / 4 = 3.14$ cm^2, $V_d = S_d * d_d = 3.14$ cm^3, respectively, and the mass is: $m_d = V_d * \rho = 18.8$ g.

We find an electric charge on the capacitor. The capacity of the flat capacitor in the practical system of coordinates is as follows:

$$C = \varepsilon\varepsilon_0 S_d / d_d = 278 \text{ pF}, \tag{1}$$

where $\varepsilon_0 = 8.85 * 10^{-12}$ F / m – dielectric permittivity of vacuum. Such a capacitor can be charged up to 280 kV [3], page 321, so that the electric charge (expressed in practical units), will be:

$$Q = CU = 7.8 * 10^{-5} \text{ Coulomb}, \tag{2}$$

i.e., the capacitor will contain: $N_e = 7.8 * 10^{-5} * 6 * 10^{18} = 4.7 * 10^{14}$ electrons.

The number of nucleons in this cylinder is equal to: $A = 1.13 * 10^{25}$ nucleons, the ratio of the charge to the mass in such a capacitor will be equal to the following: $(N_e / A) e = 4.1 * 10^{-11}$.

1.2. Acceleration of the dipoles

Similarly to the gradient of the magnetic field accelerating the magnetic dipole, the acceleration of the electric dipole is carried out by the gradient of the electric field of the wave:

$$F_e = (N_e / A) e * dE_{zw} / dz. \tag{3}$$

As for the magnetic dipole, the gradient of the electric field of the wave is $dE_{zw} / dz = k_3 * E_{zw}$, $k_3 = 2\pi/\lambda_s$, where $\lambda_s = \lambda_0 * \beta_{ph}$ - slowdown wave length in the structure, $\beta_{ph} = V_{ph} / c$ - relative phase velocity of the wave in a spiral waveguide, $c = 3 * 10^{10}$ cm / s - velocity of light in vacuum.

Finally,

$$F_e = (N_e / A) e * (2\pi d_d/\lambda_s) * E_{zw0} * \sin\varphi_s, \tag{4}$$

where E_{zw0} - the amplitude of the electric field strength on the axis of the spiral. Before substituting the numbers into formula (4), we make a few general remarks.

First of all, it should be stressed that force Fe accelerating the electric dipoles does not depend on the length of the dipole (d_d). Indeed, while decreasing d_d, the capacitance C of a the plate capacitor grows, but at the same time the breakdown voltage U decreases, thus, the charge stored in the capacitor does not depend on parameter d_d.

When parameter d_d decreases the mass of the dielectric reduces and the specific electric charge - the charge per unit of the mass, increases. However, while decreasing d_d, the accelerating force F_e acting on the dipole, decreases. This is due to the fact that the accelerating force is the difference between the repulsive force of one pole and the electric force of the other pole. From (4) it is clear that this force is greater, the greater the distance d_d is between the dipole poles.

Moreover, the accelerating force acting on the electric dipole is independent of S_d – the transverse dipole square. Indeed, with the growth of this square the charge stored in the capacitor increases, but simultaneously, in the same proportions, the mass of the dielectric also grows. This leads to the independence of the specific electrical charge of the capacitor on its transverse cross section. It is important to mention that in the case with a magnetic dipole the situation is different. The magnetic moment of the coil with the current grows as the square of the coil, i.e. the square of the increasing radius. The mass of the coil increases as the perimeter of the coil, i.e., linearly with the increasing radius, that results in linear increasing of the specific magnetic moment with the increasing radius of the coil with the current.

Now we substitute the numbers into the formula (4). The first factor (N_e / A)e determines the maximum electric charge per nucleon, which can be stored in the capacity. As it is shown above, this charge is determined only by the properties of the substance, in this case, - by the properties of the capacitor ceramic, i.e. - by the highest possible relative permittivity ε and maximum breakdown voltage U. Probably, materials with better properties will be developed later.

The second factor $2\pi d_d/\lambda s$ is determined by the ratio of the length of the dipole d_d to the slowdown wavelength λ_s in the spiral waveguide. At a large length of the accelerator it will be necessary to divide the accelerator into separate sections each of them will be individually supplied with power. Then it will be possible to accelerate the dipoles in each section at the optimal frequency for each section, which in the case of the spiral waveguide is defined by the following ratio: $\lambda_s = 2\pi r_0$, where r_0 - the radius of the spiral. The diameter of the spiral may be chosen slightly larger than d_{out} - diameter of the capacity, for example: $d_{out}/2r_0 = 0.5$, and then the parameter is equal to $d_d/r_0 = 0.5$.

Finally, for $E_{zw} = E_{zw0} * \sin\varphi_s = 250$ kV / cm, we find:

$$F_e = (N_e / A) e * (2\pi d_d/\lambda_s) E_{zw} = 5.2 * 10^{-4} \text{ eV} / (m * \text{nucleon}), \qquad (5)$$

thus, to achieve the finite velocity $V_{fin} = 8.5$ km / s,
$W_{fin} = 0.4$ eV / nucleon, the acceleration length will be required to be equal to:

$$L_{acc} = W_{fin} / F_e = 0.77 \text{ km}. \qquad (6)$$

2. The structure of the accelerator

Fig. 1 shows the scheme of the accelerator.

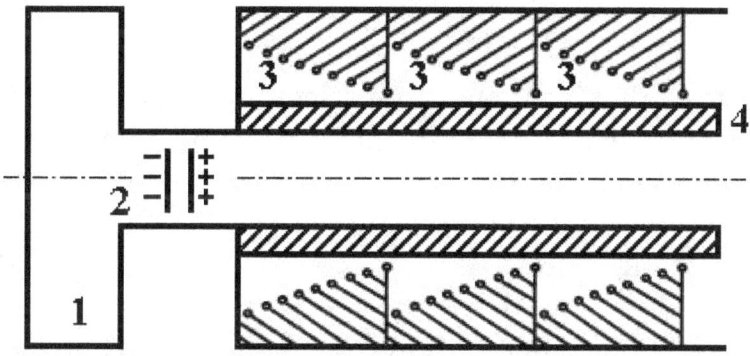

Fig. 1

Fig. 1 shows: 1 - gun, carrying out the gas-dynamic acceleration of electric dipoles, 2 - electric dipoles, pre-charged electrical capacitors, 3 - section of the spiral waveguide, 4 – dielectric channel located on the axis of the system where the acceleration of the electric dipoles is carried out.

2.1. Pre-acceleration of the electric dipoles by using the gas-dynamic method

To accelerate the electric dipoles by the field of the traveling wave, this wave must be very slow. It should be mentioned that the relative velocity $\beta = 10^{-6}$ corresponds to the normal velocity equal to: $V = 0.3$ km / s and is less than the velocity of the sound in the air. The gas-dynamic acceleration method does not allow one to achieve a higher velocity than $V_g = 2$ km / s.

For example, specifications of the gun AP 35/1000, produced by the German company "Rheinmetall" are as follows: the initial rate of shooting $V_{in} = 1.5$ km / s, the diameter of the projectile: $d_{sh} = 35$ mm. The company "Mauser" is developing an aircraft gun with a caliber (diameter of the projectile) $d_{sh} = 30 - 35$ mm and a projectile velocity $V_{in} = 1.8$ km / s.

We take the initial velocity of the electric dipoles to be reached after the gas-dynamic acceleration, equal to: $V_{in} = 1$ km / s.

2. 2. *Spiral step*

We have to take a very small step of spiral winding since we have chosen the radius of the spiral to be equal to: $r_0 = 2$ cm, then to get the spiral slowdown equal to $\beta_{ph\ in} = 3.3 * 10^{-6}$, where $\beta_{ph\ in} = V_{in} / c$ - initial phase velocity, expressed in the units of the light velocity and coinciding with the initial velocity of the electric dipoles.

The slowdown in the spiral is purely geometrical. In the simplest case the phase velocity of the wave, expressed in the units of the light velocity, in the spiral waveguide is equal to:

$$\beta_{ph} = tg\ \Psi, \tag{7}$$

where tg Ψ - the tangent of the winding angle, the tangent, in the case of large decelerations, is equal to: tg $\Psi = h/2\pi r_0$ –the ratio of the step of spiral winding to the perimeter of the coil.

Besides purely geometric slowdown of the wave in the spiral, it is possible to slow it down additionally if to place the spiral totally inside the dielectric medium with a relative permittivity ε. For barium titanate near the Curie point, it is possible to achieve the following values: $\varepsilon = 8 * 10^3$, [5], page 557. But we take a smaller value of ε: $\varepsilon = 1.28 * 10^3$, the relationship between the phase velocity of the wave propagating in the spiral and its parameters in this case, can be written in the following form [6],

$$\beta_{ph} = tg\ \Psi\ /\ \varepsilon^{\frac{1}{2}}, \tag{8}$$

The area inside the spiral should be left free of the dielectric because the acceleration of the electric dipoles will take place in this region along the axis of the spiral. Then, for the spiral where the dielectric has filled the region located between the coil and the external screen, the dispersion equation – the equation relating the parameters of the spiral with the phase velocity of the wave looks as follows, [6],

$$\beta_{ph} = \sqrt{2} * tg\ \Psi / \varepsilon^{\frac{1}{2}}. \qquad (9)$$

For the start of the spiral, where the velocity of the electric dipoles is equal to $V_{in} = 1\ km\ /\ s$, $\beta_{ph\ in} = 3.3 * 10^{-6}$, $r_0 = 2\ cm$, $\varepsilon = 1.28 * 10^3$, from (9) we find that the winding step of the spiral must be equal to:

$$h = 10^{-3}\ cm. \qquad (10)$$

The amplitude of the field strength E_{zw0}, which we have chosen, is equal to: $E_{zw0} = E_{zw} / sin\varphi_s = 350\ kV\ /\ cm$, $sin\varphi_s = 0.7$.

At step $h = 10^{-3}\ cm = 10\ \mu$ there is a risk of the electric breakdown of the dielectric. Breakdown voltage of the polyimide is 300 MV / m, [5], page 550, or 300 V / μ, thus, you can choose the spiral structure to be as follows: copper coil with a cross section of 8 μ and isolation of polyimide 2microns thick.

2. 3. The required wave power

Relationship between the power flux and the wave strength on the spiral is given by the following formula [6],

$$P = (c/8)E_{zw0}^2 r_0^2 \beta_{ph}\{(1+I_0K_1/I_1K_0)(I_1^2 - I_0I_2) +$$
$$\varepsilon(I_0/K_0)^2(1+I_1K_0/I_0K_1)(K_0K_2 - K_1^2)\}. \qquad (11)$$

The argument of the modified Bessel functions of the first and second types presented in the curly brackets is the value of $x = 2\pi r_0/\lambda s$, which we have chosen to be equal to: $x = 1$. Then, for this argument the second term in curly brackets is much greater than the first term, and the curly bracket itself is equal to: $\{\} = 3.77 * \varepsilon$. Substituting the numbers into the formula (11), $\varepsilon = 1280$, we find

$$P\ (W) = 3 * 10^{10} * 3.5 * 3.5 * 10^{10} * 4 * 3.3 * 10^{-6} * 3.77 * 1.28 * 10^3 /$$

$/ (8 * 300 * 300 * 10^7) = 125$ MW.

In order to reach the field strength at the axis of the spiral to be equal to $E_{zw0} = 350$ kV / cm, it is required to introduce power $P = 125$ MW.

This power can be achieved by the pulse technology.

We expand the sinusoidal pulse [6], corresponding to the half-wave $E_{pulse} = E_{0pulse} \sin (2\pi/T_0)\, t$, $2\pi/T_0 = \omega_0$, $\omega_0 = 2\pi f_0$ in a Fourier row:

$$f_1 (\omega) = (2 / \pi)^{1/2} \int_0^{T_0/2} \sin\omega_0 t * \sin(\omega t)dt. \qquad (12)$$

The pulse spectrum is rather narrow and covers the frequency range from 0 to $2\omega_0$. Since the spiral waveguide dispersion (in dependence of the phase velocity on frequency) is weak, it can be expected that the full range of frequencies from 0 to $2\omega_0$ will propagate approximately with the same phase velocity.

As a result, the half-wave sinusoidal pulse in vacuum will spread out only due to increasing of the phase velocity of the wave. In this case matching of the spiral waveguide with a power feeder is necessary to carry out in the frequency band: $\Delta f \approx \omega_0/2\pi$.

We introduce the notion of the pulse amplitude U_{acc}, related with the field strength at the axis of the spiral E_{zw0} by the following ratio [6]:

$$U_{acc} = E_{zw0}\lambda_s/2\pi, \quad \lambda_s = \beta\lambda_0, \quad \lambda_0 = c/f_0. \qquad (13)$$

In this case, the vacuum wavelength λ_0 is: $\lambda_0 = \lambda_s / \beta_{in} = 4.18 * 10^6$ cm, the wave frequency: $f_0 = c/f_0 = 7.1 * 10^3$ Hz, half-life - the duration of the pulse on the basis is equal to: $T_0 / 2 = 1/2f_0 = 90$ μs.

Thus, the amplitude of the voltage pulse propagating along the spiral must be equal to: $U_{acc} = E_{zw0} * \lambda_s/2\pi = 700$ kV. The Table below summarizes the main parameters of the accelerator.

Table. Parameters of the accelerator.

Parameter	Value
Number of electrons per nucleon in the electric dipole, $(Z/A)e$	$(Z/A)e = 4.1*10^{-11}$
The ratio of the dipole to the slowdown-wave length $2\pi d_d /\lambda_s$	$2\pi d_d /\lambda_s = 0.5$
Wave power in Watts, P	$P = 125$ MW
The speed of the electric dipoles, the initial - final, $\beta = \beta_{ph}$	$\beta_{ph} = 3.3*10^{-6} - 2.83*10^{-5}$
The radius of the spiral, r_0	$r_0 = 2$ cm
The frequency of the wave, f_0,	$f_0 = 7.1*10^3$ Hz
The tension of the electric. field E_{zw0}	$E_{zw0} = 350$ kV/cm
The length of the accelerator, L_{acc}	$L_{acc} = 0.77$ km
Pulse duration, τ	$\tau = 90$ μs
The amplitude of the voltage, U_{acc}	$U_{acc} = 700$ kV

2. 4. Phase stability

It is known that in the traveling wave the phase stability region is on the front slope of the wave pulse. Indeed, if the particle is faster than the wave, it will get into the weakening field, and, finally, the speeding up pulse will soon catch up with the particle.

If the particle is behind its phase, it will get into the region of the strengthening field and acquire more energy in comparison with what would have been obtained in the synchronous phase and, eventually, the particle will catch up with its phase.

Thus, from the viewpoint of the mutual position of the particle and the acceleration pulse there is only possible case when the pulse «pushes» but not «pulls» the particle. In our calculations we have chosen the synchronous phase to be equal to: $\varphi_s = 45^0$, $\sin \varphi_s = 0.7$. To achieve a greater rate of acceleration, it is possible to choose a greater value $\varphi_s = 60^0$, $\sin \varphi_s = 0.87$, but then at the acceleration it will be needed to meet more strict requirements.

It is known that while acceleration of particles in azimuthal - symmetrical field, the phase stability corresponds to the radial instability. This means that when we push the particle by the pulse, we push it not only forward but also a bit on the radius. The force acting on the particle increases linearly with the growth of particle deviation from the axis and the initial deviation increases exponentially.

To keep the particles near the axis, it is necessary to introduce focusing while accelerating, i.e., it is necessary to use additional forces, fields which will return the particle to the axis of the acceleration.

2. 5. Preventing the turn of the dipole by 180^0 in the electric field of the pulse and keeping it on the axis of the dipole acceleration

In the acceleration of the dipoles there is a new problem which did not exist while accelerating the point particles. The pulse accelerating the dipole will lead to a roll-over of the dipole - turn it by 180 degrees.

This problem is easier to see with the example of the acceleration of the magnetic dipoles and conventional magnets by the running current pulse. If such a pulse, pushing the magnet, to substitute by another magnet, it can be seen that the magnet being pushed will not be easy to push from the same sign pole of the pushing magnet but it will try to turn by 180 degrees and pull itself to the opposite-sign pole.

The action of the pair of forces leading to the "reversal" of the dipole is summed up in comparison with the "difference" forces accelerating the dipole and leading to the radial displacement of the center of mass.

The simplest solution that prevents reversal of the magnetic dipoles by 180 degrees in the accelerating wave field is the imposition of the uniform external magnetic field. It won't influence the acceleration of the dipole because of its homogeneity but will only hold the dipole against the reversal in the space.

By analogy with the uniform magnetic field which does not affect the acceleration of the dipoles but keeps the magnetic dipoles against the reversal by 180 degrees, it is possible to use the uniform electric field to hold the electric dipole against the reversal.

The intensity of the electric field must be at least of a larger wave amplitude, i.e. $E_{keep} > 350$ kV / cm. When acceleration length L_{acc} is ≈ 1 km, the electrostatic field for this purpose is not acceptable since in this case a tremendous difference of potentials would take place.

The uniform electric field can be formed by the induction method. Let the length of the inductor be (along the acceleration axis) $l_{ind} = 10$ cm. Such a length will be covered by the electric dipole with velocity $V_{in} = 1$ km / s during the period of time equal to $\tau_{ind} = 10^{-4}$ s, hence, the induction method should create tension $U_{ind} = E_{keep} * l_{ind} = 3.5$ MV for the period of time $\tau_{ind} = 10^{-4}$ s.

Multiplying these values, we find that the required amount of the magnetic flux is $F = 350$ T $* $ m^2. Let this flux be created by iron re-magnetization from the value of -1 T till the value of +1 T, then the cross-section of the iron in the inductor should be $S_{ind} = 175$ m^2. Then at the length of the inductor along the acceleration axis - $l_{ind} = 10$ cm, its radial length should be: $r_{ind} = 1750$ m, that is not justified either.

It is possible to hold the electric dipoles near the acceleration and at the same time - to keep them against turning by 180 degrees, if you keep the acceleration of the electric dipoles in a narrow dielectric channel, located on the axis of the system. In this case at the longitudinal acceleration of the electric dipoles in the channel the dipole will undergo friction with the walls of the channel.

2. 6. The influence of friction

We consider how strong the friction of the side surface of the electric dipoles over the internal surface of the channel, will influence the acceleration.

The side surface of the electric dipoles must be dielectric to it avoid the short current in the capacitor. The channel must be also dielectric not to screen the electric field accelerating the electric dipoles along the spiral structure.

Actually, the friction force directed against the accelerating force is equal to the coefficient of friction multiplied by the radial force pressing the electric dipoles towards the inner surface of the channel. In its turn, the radial force proportional to the acceleration force and deviation of the electric dipole from the axis is absent when the center of the electric dipoles is located on the axis of the system.

Let the deviation of the electric dipoles from the axis of the system is 1% of its radius, $d_{out} / 2 = 1$ cm, i.e. $\Delta r = 100$ μ. From the expansion of the Bessel function of the first order with a small argument, it is clear that a pair of forces acting on a dipole is equal to the following:

$$F_r = (2\pi\Delta r/\lambda_s) (N_e / A) e * (2\pi d_d/\lambda_s) E_{zw}. \qquad (14)$$

A typical value of the coefficient of friction, for example, by cellophane [5], page 128, is: $k_{fr} = 0.4$. On the one hand, the coefficient of friction increases when the friction surfaces are in vacuum, on the other hand, it decreases with increasing of the relative velocity of motion. We assume that for our velocity of the electric dipoles moving within the channel, the friction coefficient is about the same value. Then, the ratio of force directed against velocity, F_f, to the force of acceleration is equal to

$$F_f / F_e = k_{fr} * (2\pi\Delta r/\lambda_s) = 2 * 10^{-2}, \qquad (15)$$

that will be compensated by phase stability.

According to Hooke's law we find the elastic force F_{elas}, which will prevent the compression of the electric dipoles:

$$\Delta r / d_{out} = F_{elas} / (E * S_c), \qquad (16)$$

where: $\Delta r / d_{out} = 5 * 10^{-3}$ - relative compression, E - Young's modulus, S_c - the contact surface.

At a height of segment Δr, the length of chord a can be found from the

70

formula:

$$a = 2 (\Delta r * d_{out} - \Delta r^2)^{1/2} \approx 2 (\Delta r * d_{out})^{1/2}, \qquad (17)$$

and, if the contact surface is: $S_c = d_d * a = 3 * 10^{-5}$ m². A typical value of Young's modulus E for plastics is equal to $E = 10^8$ N/m², [3], p.53, and from (16) we find that the value of force F_{elas} is: $F_{elas} = 15$ N, that is comparable with radial force Fr which must be multiplied by the number of nucleons $A = 1.13 * 10^{25}$, $F_r = 5*10^{-3}*5.2*10^{-4}*1.6*10^{-19}*1.13*10^{25} = 5$ N.

It is clear that the elastic force returning the center of gravity of the electric dipoles onto the axis of the system will prevent a significant radial displacement of the center of gravity of the electric dipoles relatively the axis of the system.

Conclusions

The efficiency of the acceleration of charged cylindrical bodies decreases with increasing of the diameter of the cylinder. This is due to the fact that the electric charge is located on the cylinder surface. The square of the unit of the cylinder length increases as the radius of the cylinder, and the volume and mass increase as the square of the radius. As a result, the charge per unit of the (nucleon) mass reduces with increasing of the radius as 1 / r. From a certain radius the acceleration rate of the electric dipoles by the wave field gradient becomes greater than the acceleration rate of the charged bodies.

The choice of electric dipoles in the form of a plate capacitor allows the electric dipoles to reach the first order space velocity at a distance less than one kilometer.

Literature

1. S. N. Dolya, Methods of acceleration of the magnetic dipoles, Patent of the Russian Federation, № 2451894

2. S. N. Dolya, Method of acceleration of the macro particles, Patent of the Russian Federation, № 2456782

3. Tables of physical quantities, Directory Ed. I. K. Kikoin, Moscow, Atomizdat, 1976

4. http://ru.wikipedia.org/wiki/Титанат_бария
   http://ru.wikipedia.org/wiki/Титанат_стронция

5. Physical quantities, Directory Ed. I. S. Grigor'ev and E. Z. Meylikhov, Moscow, Nuclear Power publisher house, 1991

6. S.N. Dolya, K.A. Reshetnikova, On the electrodynamics' acceleration of macroscopic particles, Communication JINR P9-2009-110, Dubna, 2009, http://www1.jinr.ru/Preprints/2009/110(P9-2009-110).pdf
http://arxiv.org/ftp/arxiv/papers/0908/0908.0795.pdf

III. Application accelerators

On the implementation of the conditions of Inertial Confinement Fusion by bombarding the target a macro particle

The acceleration of lithium tube segments with the length l_s = 1 cm, diameter d_s = 16 μ, wall thickness δ_s = 1 nm up to the energy W_{fin} = 1 MeV / nucl is considered. These segments are electrically charged up to the surface field strength E_s = 10^9 V / cm by proton beams produced by an electron beam source, which results in a charge-to-mass ratio $Z / A \approx 1.6 * 10^{-2}$. Then, they are accelerated by the traveling wave field in a spiral waveguide at the length of acceleration $L_{acc} \approx 30$ m. The segments are next sent to a frozen (D, T) target where they are compressed by three hundred times in the longitudinal direction while compressing the (D, T) target radially by 10^4 times—thus, the conditions for thermonuclear fusion are realized.

1. Introduction

A prerequisite of thermonuclear burning in a reactor with inertial-confinement fusion (ICF) is the target compression by a power flux from external radiation sources. This is done in order to increase by thousands times the reaction yield $Y = \sigma n l_t$, where σ is the reaction cross section; n is the density of the target atoms; l_t is the interaction length which is about the target size. While compressing a spherical target, its density increases as $1/r^3$, and the interaction length decreases as r, so that the product $n l_t$ may grow significantly, and the yield will increase from the normal for nuclear reactions values $Y \sim 10^{-4}$ to the value $Y \sim 1$ as the thermonuclear fuel is burned completely.

72

The well-known schemes of compression by laser radiation and light ion beams do not allow (under focusing conditions) compression of targets with a characteristic transverse dimension of several microns. However, no such limitation exists for the compression of targets by macro particles accelerated to energies above the threshold of nuclear reactions.

This study examines the acceleration of macro particles in a spiral waveguide and processes occuring as they hit a (D, T) target.

2. The parameters of an accelerated particle

Let us consider a macro particle representing a thin-wall lithium tube section, namely, a tube segment with the diameter $d_s = 16 \, \mu$, wall thickness $\delta_s = 10^{-7}$ cm, (1 nm), and length $l_s = 1$ cm. The mass of the tube section is: $M_s = \rho_{Li} * \pi \, d_s \, \delta_s \, l_s \approx 2.6 * 10^{-10}$ g, where the density of lithium is taken to be $\rho_{Li} = 0.53$ g/cm^3.

We find from Avogadro's ratio that the number of atoms per cubic centimeter of lithium is $4.5 * 10^{22}$ atoms. A tube segment with the specified parameters contains $N_{Li} = 2.25 * 10^{13}$ atoms of lithium or approximately $A = 1.6 * 10^{14}$ nucleons.

When irradiating such a macro particle by an ion beam, it can be electrically charged up to the surface field strength $E_s = 10^9$ V / cm. To be specific, we will refer here to a proton beam.

The density of the field electron emission current is j (A/cm^2) = 10^{-4} for the surface barrier $e\varphi = 2.38$ eV (electronic work function for lithium, [1], p. 444) and the surface field intensity $E_s = 10$ MV / cm, [1], p. 461.

Let us assume that the surface intensity of the electric field on a tube segment is $E_s = 10^9$ V / cm, and the density of the leakage current of positive charges from the tube section equals j = 10^{-4} A/cm^2.

The charge density κ located on the cylindrical surface is related to the surface intensity of the field E_s by the relationship:

$$E_s = 4\kappa/d_s, \qquad (1)$$

does not depend on the cylindrical surface length, and is determined only by its diameter.

We find from (1) that at the surface field strength $E_s = 10^9$ V/cm one-centimeter of the tube section length contains $N_p = 2.6 * 10^{12}$ uncompensated protons.

Let us calculate the ratio of charge per nucleon in such a cylinder. To do so, we divide the total number of excess protons, $N_p = 2.6 * 10^{12}$, placed on a tube segment by the total number of nucleons contained in this segment, $A = 1.6 * 10^{14}$. We obtain that this ratio $N_p / A = 1.6 * 10^{-2}$, which roughly corresponds to a fourfold ionized ion of uranium 238 ($4/238 = 1.7 * 10^{-2}$).

The energy of ions irradiating the macro particle should vary from the primary, equal to units of electron-volts, to the end one:
$e\varphi = (½) E_s * d_s = 800$ keV.

A short explanation should be given as to how such a lithium thin-wall tube segment can be made. It can be produced by spraying a few layers of lithium (10 Angstroms = 1 nm) onto the inner surface of any tube-shaped shell, for instance, from an organic material. Obviously, before such a tube segment is further irradiated to provide its positive charge, this shell should be removed, for instance, through evaporation under laser irradiation.

The head and tail parts of the tube section should be closed with lithium hemispheres of the same wall thickness with the result that it must have the shape of a vessel operating under pressure.

This segment of thin-wall lithium tube with closed hemispheres of the head and tail parts is, in fact, a lithium bubble.

3. Acceleration of particles

We consider the acceleration of macro particles which can be carried out according to the usual scheme: preliminary electrostatic acceleration up to a speed that approximately coincides with the phase velocity of the wave in a slowing structure and final acceleration in the traveling wave field.

3. 1. Static field acceleration

Let us assume that the voltage under which the platform with the container comprising tube segments is kept equals U = 800 kV. Then, after electrostatic acceleration the relative speed of the segment, $\beta = V / c$, expressed in terms of the speed of light in vacuum $c = 3 \times 10^{10}$ cm / s, will be:

$$\beta = [2e \, (Z / A) \, U/Mc^2]^{1/2}, \qquad (2)$$

where $Mc^2 = 1$ GeV is the rest mass of the nucleon expressed in units of electron-volt; $Z / A = 1.6 * 10^{-2}$ is the electric charge per nucleon; and e is the unit charge. Substituting the numbers into formula (2), we find that the speed of macro particles after preliminary acceleration equals $\beta = 5 * 10^{-3}$. The initial velocity β_{in} of the wave traveling in a slowing structure- a wave whose longitudinal electric field component is bound to accelerate macro particles - should be the same.

3. 2. Electromagnetic acceleration

3. 2. 1. The parameters of a spiral waveguide

We choose a spiral waveguide as an accelerating structure for the acceleration of macro particles, [2].

For a spiral waveguide with the spiral winding radius $r_0 = 1$ cm, the relationship β between the winding pitch and phase velocity therein can be obtained:

$$\beta = \sqrt{2} * tg\Psi/\varepsilon^{1/2}, \qquad (3)$$

where $tg\Psi = h/2\pi r_0$ is the tangent of the winding angle; h is the spiral pitch; $2\pi r_0$ is the turn perimeter; ε is the dielectric permittivity factor of the medium filling the space between the spiral and the outer shield.

Let us take water as a filler of the space between the spiral and the outer shield: $\varepsilon = 80$, $\varepsilon^{1/2} \approx 9$. Then, for the spiral radius $r_0 = 1$ cm and the initial velocity of the wave in the slowing structure $\beta = \beta_{in} = 5 * 10^{-3}$, we find from relation (3) that $h_{in} = 0.28$ cm.

In a spiral, waves of different frequencies propagate with the same deceleration, but there is a frequency f_0 for which the maximum strength of the longitudinal field E_z can be achieved at a given flow rate. This frequency may be determined from the formula:

$$2\pi r_0/\beta\lambda_0 = 1, \qquad\qquad (4)$$

where $\lambda_0 = c/f_0$ is the vacuum wavelength. From (4) we find the optimal wavelength of acceleration (for the beginning of the accelerator): $\lambda_{0in} = 12.5$ m. The corresponding frequency $f_{0in} = c/\lambda_{0in} = 24$ MHz.

We now choose an end-point energy up to which such lithium macro particles will be accelerated: $W_{fin} = 1$ MeV / nucl. Let us assume that the voltage to accelerate a micro particle $E_0\sin\varphi_s = 20$ kV / cm, where $E_0 = 30$ kV / cm is the field amplitude; $\varphi_s = 45^0$ is the synchronous phase, $\sin\varphi_s = 0.7$.

Knowing the effective charge of a macro particle $Z / A = 1.6 * 10^{-2}$ and intensity of the accelerating field $E_0\sin\varphi_s = 20$ kV / cm, you can find the accelerator length L_{acc2} from the relationship:

$$W_{fin} = (Z / A)\ e\ E_0\sin\varphi_s * L_{acc2}. \qquad\qquad (5)$$

Hence: $L_{acc2} = W_{fin} / [(Z / A)\ e\ E_0\sin\varphi_s] \approx 30$ m.

Such a large accelerator length is due to the fact that the amount of charge per nucleon has a very low value for a macro particle, $(Z/A) = 1.6 * 10^{-2}$, while being approximately $(Z/A) = \frac{1}{2}$ for light ions, and for protons equal to unity.

3. 2. 2. Power consumption

We shall find the power required to create in a spiral waveguide the field strength $E_0 = 30$ kV / cm for the start of the accelerator from the relationship, [2]:

$$P_{in} = (c / 8)\ E_0^2 * r_0^2 * \beta_{in} * \{(1 + I_0K_1/I_1K_0)\ (I_1^2 - I_0I_2) + \varepsilon\ (I_0/K_0)^2\ (1 + I_1K_0/I_0K_1)\ (K_0K_2 - K_1^2)\}, \qquad (6)$$

where I_0, I_1, I_2 are modified Bessel functions of the first kind, and K_0, K_1, K_2 are modified Bessel functions of the second kind. The first term in the curly brackets corresponds to the flux propagating inside the spiral, while the second term to the flux traveling between the spiral and external shield. The latter area is filled with a medium having a dielectric constant ε; therefore, the second summand contains a factor ε.

The Bessel functions in (6) are dependent on the argument $x = 2\pi r_0/\beta\lambda_0$. Where the spiral is most effective, $x = 1$, the first term in the curly brackets is equal to unity and the second term to 4 ε.

In our case, $\varepsilon = 80$, the first term can be neglected, and substituting the numbers into formula (6), we obtain:

$$P_{in} = 3 * 10^{10} * 9 * 10^8 * 5 * 10^{-3} * 4 * 80 / (8 * 9 * 10^4 * 10^7) = 6 \text{ MW}.$$

Let us determine the required power for the end of the accelerator, assuming that the same spiral radius $r_0 = 1$ cm is taken. Since the phase velocity has increased by two orders towards the end of the accelerator, there is no water filling of the space between the spiral and the outer shield. Then the power P_{fin} equals:

$$P_{fin} = 3 * 10^{10} * 9 * 10^8 * 4.5 * 10^{-2} * 4 / (8 * 9 * 10^4 * 10^7) = 675 \text{ kW}.$$

We can see that obtaining such power for the pulsed operation of radio frequency generators will not be a problem.

The required high-frequency power at the accelerator's beginning is greater compared to its end, which is due to the use of a dielectric between the spiral and outer shield. According to formula (6), only 0.3% of high-frequency power propagates along the spiral axis where the lithium tube segments are accelerated. This use of the dielectric is explained by the necessity to increase the spiral pitch at the beginning of the accelerator. Let us find the spiral pitch h_{fin} at the end of the accelerator from the relation $\beta_{fin} = h_{fin}/2\pi r_0$. Substituting $r_0 = 1$ cm and $h_{fin} = 2.8$ mm, we obtain: $\beta_{fin} = 4.5 * 10^{-2}$.

3. 2. 3. The capture of macro particles into the acceleration mode. Tolerances.

We calculate now the accuracy which is required for the initial phase of the accelerating wave to coincide with the synchronous phase. The theory of particle capture by a traveling wave shows [3]: $\Delta\varphi = 3\varphi_s, (+ \varphi_s, - 2\varphi_s)$. In our case, the quarter period $1/4f_0 = T_0/4$ corresponds to 10 ns or 90^0; hence, one degree in phase corresponds to a time interval of about 0. 1 ns.

A buncher in a linear accelerator provides a bunch phase width of $\pm 15^0$. In

order to avoid large phase fluctuations, the accuracy of synchronization between the macro particle and accelerating wave should be:
$\Delta\tau = \pm 15 * 0.1\text{ns} = \pm 1.5$ ns.

Such timing accuracy is apparently quite achievable.

Let us calculate the tolerance for the accuracy of matching the initial speed of a macro particle and phase velocity of the pulse propagating along the spiral structure. We introduce the value $g = (p-p_s) / p_s$, which is a relative difference in pulses, [3]. In a non-relativistic case, this corresponds to a relative velocity spread: $g = (V-V_s) / V_s$. The vertical spread of the separatrix is calculated by the formula, [3]:

$$g_{max} = \pm 2 \ [(W_\lambda \text{ctg}\varphi_s/2\pi\beta_s) * (1 - \varphi_s / \text{ctg}\varphi_s)]^{1/2}, \qquad (7)$$

where $\varphi_s = 45^0 = \pi / 4$, $\text{ctg}\varphi_s = 1$, $[1 - \varphi_s / \text{ctg}\varphi_s]^{1/2} = 0.46$, $2 * 0.46 = 0.9$
$W_\lambda = (Z / A) \ eE_0\lambda_0\sin\varphi_s/Mc^2$.

Further, we determine the value $W_\lambda = (Z / A) \ eE_0\lambda_0\sin\varphi_s/Mc^2$, which is a relative energy gain by a macro particle at the wavelength λ_0 in vacuum. In our case: $\lambda_0 = c/f_0 = 12.5$ m, $\sin\varphi_s = 0.7$, $Mc^2 = 1$ GeV, $W_\lambda = 4 * 10^{-4}$. Substituting numerical values, we obtain that $g = (V_{in}-V_s) / V_s = \Delta V / V_s$, and, finally,
$\Delta V/V_s = \pm [4*10^{-4}/ (6.28*5*10^{-3})]^{1/2}*0.9 = \pm 0.1$.

Thus, the admissible discrepancy between the initial speed of the macro particle and the wave velocity is of the order: $\Delta V / V_s = \pm 10\%$.

3. 2. 4. *The heating of the spiral*

In the above, we have found high-frequency power losses associated with attenuation. These losses are spent for heating of the spiral, and in order to prevent changing of its electromagnetic characteristics and melting, the spiral needs to be cooled. This can be done with water, which acts as a medium with a large dielectric constant and is located between the spiral and external shield.

At the end of the accelerator where there is no such medium, cooling needs to be done by gas such as helium.

3. 2. 5. *Radial movement*

As is well known, [3], in azimuth-symmetric waves the phase stability (or autophasing) region is characterized by radial defocusing. Therefore, you cannot obtain radial and phase stability in such waves concurrently. However, when the phase stability is achieved, radial stability can be provided as well by introducing external fields. In this phase region, the radial component of the electric field of the wave is directed towards radius increase, accelerating thus the particles in the direction away from the axis of acceleration.

Let us consider focusing of accelerated lithium tube segments using quadrupole lenses. It is known, [3], that quadrupole lenses simultaneously provide focusing of particles in one plane and defocusing in another. If two lenses are used so that they are turned relative to each other through 90^0, then focusing and defocusing areas will be generated alternately in each of the transverse planes. Under certain conditions, such a system of lenses appears to be focusing one. The doublets are usually placed between adjacent sections being turned through 90^0.

In fact, a particle moving accurately along the axis is not acted on by any forces. The further away the particle is from the axis, the greater is the action of the forces. Let the particle hit the focusing area. Its trajectory will bend then in such a way as to cross the defocusing area at the minimal value of the field, and the focusing forces will prove to be greater than defocusing ones. A similar effect also arises if the particle passes the defocusing area first. The resulting effect from a pair of quadrupole lenses will be collecting, [3].

The focusing and defocusing effects of the lenses are determined by their rigidity:

$$K = [(Z / A) eGl_1^2/Mc^2\beta_z], \qquad (8)$$

where (Z / A) is the ratio of charge to mass; G is the gradient of the electric or magnetic field in the lens; l_1 is the length of the lens; $Mc^2 = 1$ GeV corresponds to the rest mass of the nucleon; β_z is the longitudinal particle velocity expressed in terms of the speed of light. Rigidity among these notations is a dimensionless quantity.

In contrast to a pair of quadrupole lenses, the accelerating section operates as a defocusing lens in both perpendicular directions. Around the acceleration axis where there is no electric volume charge the following condition is realized:

$$\text{div } E = 0, \qquad\qquad (9)$$

from whence follows the ratio between the longitudinal and transverse electric fields:

$$E_r = - (r / 2) \, dE_z / dz, \qquad\qquad (10)$$

which is apparent, however, from the structure of the field in the spiral waveguide, [2].

By analogy with quadrupole lenses, we can introduce for the accelerating field the concept of the field gradient $G_s = \frac{1}{2} \, dE_z / dz = \pi E_0 / \lambda_s$, where E_0 designates the amplitude of the accelerating field; λ_s is the length of the slow wave in the structure.

Let us consider now the focusing by electrostatic quadrupole lenses. We require that the rigidity of the quadrupole lens be greater than the rigidity of the accelerating section, which means that the particle deflection angle in the lens must be greater than the deflection angle in the section. In the accelerating section, the angle of particle deflection is always directed outwards and defocusing of the accelerated particles occurs in both transverse planes. The particles passing through the quadrupole lens are deflected inwards while being deflected outwards in a different plane. [2].

This means that the gradient of the field in the lens multiplied by the square of its length must be greater than the gradient of the field in the accelerating section, also multiplied by the square of its length:

$$G_l * l_l^2 > G_s * l_s^2. \qquad\qquad (11)$$

The most difficult conditions in terms of focusing are found at the beginning of the accelerator.

Let the length of the lens make up one-third of the section length, that is $l_s^2 / l_l^2 = 10$, $l_s = 3 l_l$. Then, the electric field gradient in the electrostatic lenses must exceed at least by ten times the electric field gradient in the sections, i.e:

$$G_l > 10 \, G_s. \qquad\qquad (12)$$

Substituting the numbers for the start of acceleration, where $\lambda_s = 6.28$ cm, we find that the electric field gradient in the electrostatic lenses must be greater than: $G_l > 1.5 * 10^5$ V/cm^2.

Such high gradients in electrostatic quadrupole lenses are the "payment" for the high and continuous (for a traveling wave in lengthy sections) rate of acceleration. If such field gradients in the lenses are difficult enough to obtain, you will need to move to a lower rate of acceleration.

It needs to be added that such deceleration rates as $\beta_{in} = 5 * 10^{-3}$ can be easily achieved in accelerators with an azimuthally asymmetric accelerating field, the so-called RFQ (Radio Frequency Quadrupole) accelerators where focusing is performed by the accelerating field and no other focusing elements are necessary.

4. The power transmitted to the beam

We have calculated the RF power of generators needed to generate an electric field of intensity $E_0 = 30$ kV / cm in a spiral waveguide. Let us now estimate the power transferred to the beam.

Each macro particle contains $N_{Li} = 2.25 * 10^{13}$ lithium atoms with the energy $W_{Li} = 7$ MeV. Multiplying these figures, we find the energy of a macro particle in electron volts; and dividing the resulting expression by $6.24 * 10^{18}$, we obtain the energy of a macro particle in Joules: $\mathcal{E}_{Li} = 25$ J.

During the pulse operation of high-frequency generators the number of such particles in the wavetrain is $2 * 10^3$ at the pulse duration $\tau_{pulse} = 100$ μs, so that the power delivered to the beam from a generator is found to be $P_{beam} = 500$ MW, which indicates that the high frequency power of the generators must be increased, by about 100 times at the beginning of the accelerator.

It can be seen that the energy transferred to the beam is significantly greater than the energy expended on the generation of the accelerating field. If high-frequency energy is recuperated, the efficiency of the accelerator can be raised even higher.

Such power transmitted to the beam, $P_{beam} = 500$ MW, is too great for the modern high-frequency generators.

81

Standing-wave linear (i.e. RFQ) accelerators can be operated using stored energy. In order to reduce the power transmitted to the beam up to 5 MW a lithium tube segment needs to be placed into every hundredth separatrix.

In traveling-wave linear accelerators, the acceleration of particles can be provided by a faster running powerful single pulse instead of a high frequency wave [2], which ensures a corresponding growth of power.

Let us analyze the sine-shaped pulse with the help of the Fourier integral: $E = E_0 \sin(2\pi/T_0)t$, where $2\pi/T_0 = \omega_0$, $\omega_0 = 2\pi f_0$. We obtain:

$$f_1(\omega) = (2/\pi)^{1/2} \int_0^{T_0/2} \sin\omega_0 t * \sin\omega t dt. \qquad (13)$$

The pulse spectrum is narrow and occupies a frequency range from 0 to $2\omega_0$. The dispersion (dependence of the phase velocity on the frequency) in a spiral waveguide is weak, and the waves within this frequency range are expected to propagate with the same phase velocity. As a result, the half-wave sinusoidal pulse will spread out in space due to increasing phase velocity of the wave. Matching the spiral waveguide with the supply system must be implemented in this case in the frequency range: $\Delta f \approx \omega_0/2\pi$.

We introduce the concept of pulse amplitude U_{acc} associated with the field on the spiral axis E_{0pulse} using relation [2]:

$$U_{acc} = E_{0pulse}\lambda_{slow}/2\pi, \quad \lambda_{slow} = \beta\lambda_0, \quad \lambda_0 = c/f_0. \qquad (14)$$

As a result, the amplitude of the voltage pulse propagating along the spiral axis is as follows: $U_{acc} = E_{0pulse} * \lambda_{slow}/2\pi = 30$ kV. The amplitude of the current pulse $I_{acc} = P / U_{acc} = 17$ kA; the impedance line $\rho_{line} = U_{acc} / I_{acc} = 2$ Ohms.

Such a low resistance can be obtained in an artificial transmission-line with lumped parameters where the spiral winding appears as a distributed inductance (640 nH / cm) - the winding must be respectively
capacity - loaded (160 nF / cm).

Thus, we have obtained a multi-section coil gun. The difference between ours and the known samples is that we are dealing with the acceleration of particles having characteristic transverse dimensions of the order of several microns. Furthermore, we have chosen the particles in the form of a bubble where the entire mass is concentrated near the surface, and to reduce the field emission we have charged the bubble with positive particles. As is known, the field ion emission arises when the field intensity begins to significantly exceed the field electron emission. This allows one to increase the charge-to-mass ratio of the particles and accelerate them to the energy above the threshold of nuclear reactions at a length of some tens of meters.

Table of parameters

The accelerated lithium tube segment: length, diameter, wall thickness	$l_s = 1$ cm, $d_s = 16$ μ, $\delta_s = 1$nm
The number of atoms in the segment, the number of nucleons	$N_{Li} = 2.25*10^{13}$, $A = 1.6*10^{14}$
The number of protons placed on the segment	$N_p = 2.6*10^{12}$
The charge per nucleon in the interval	$Z/A = 1.6*10^{-2}$
The energy of protons irradiating the cylinder, initial—final	$e\varphi = 10$ eV - 800 keV
The potential difference relative to the ground beneath which is a high-voltage platform	$U = 800$ kV
The relative velocity of the segments after the electrostatic acceleration, the initial speed of the electrodynamic acceleration	$V_{in}/c = \beta_{in} = 5*10^{-3}$
The radius of the spiral at the beginning and end of the acceleration	$r_{0in} = r_{0fin} = 1$ cm
The pitch of the spiral at the beginning and end of the acceleration	$h_{in} = h_{fin} = 2.8$ mm
The frequency of acceleration at the beginning and end of the acceleration	$f_{in} = 24$ MHz, $f_{fin} = 215$ MHz
High-frequency power spent on the generation of electric field in the structure at the beginning and end of the accelerator	$P_{in} = 6$ MW, $P_{fin} = 615$ kW
The relative end speed of the lithium tube segments	$V_{fin}/c = \beta_{in} = 4.5*10^{-2}$

The acceleration length $L_{acc} = 30$ m
The power transmitted to the beam $P_{beam} = 500$ MW
The energy of each tube segment $\epsilon_{Li} = 25$ J

Figure 1 shows a scheme of the device.

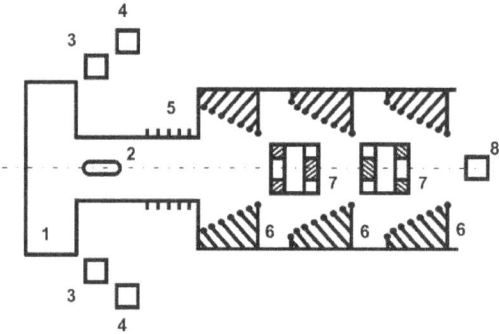

Fig.1. (1) - a container, (2) - lithium tube segments, (3) – a system of lasers, (4) - electron-beam sources, (5) – a high-voltage tube, (6) - sections of the spiral waveguide, (7) – quadrupole doublet lenses, (8) - deuterium-tritium target.

5. Thermonuclear reactions

Let us consider nuclear reactions which are to occur when such a lithium "bubble" will hit, for example, a deuterium-tritium target.

The density of liquid hydrogen at the temperature -260 ^{0} C is $\rho = 0.076$ g/cm^3; the density of deuterium is twofold higher; and of tritium three times greater, [1], p. 57. A mixture of deuterium and tritium, 50% to 50%, will have a molecular weight of 5 g and density approximately equal to 0.2 g/cm^3.

We find the number of molecules contained in one cubic centimeter from Avogadro's relationship whence one cubic centimeter contains $2.4 * 10^{22}$ molecules or about $5 * 10^{22}$ atoms.

Next, we find the depth of the path of lithium nuclei with an energy of 1 MeV / nucleon in a deuterium-tritium mixture.

The stopping power of protons in the air, at an energy of protons 1 MeV, is equal to $dW / dx = 150$ MeV $* cm^2 / g$, [1], p. 953. In most of the path the losses are equal to $dW / dx = 600$ MeV $* cm^2 / g$.

The stopping power of hydrogen in air is about two and a half times greater, [1], p. 953, and fully ionized (due to stripping) lithium ions will have the same path as protons. Altogether, one can assume that the stopping power of the deuterium-tritium mixture for fully ionized lithium ions equals: $dW / dx = 1.5$ GeV $* cm^2 / g$. Multiplying the stopping power by the density of the mixture, we obtain the energy losses through ionization $dW / dx = 300$ MeV / cm and the range of the lithium ions l_{fr}: $W_{fin} / (dW / dx) = l_{fr} = 30$ μ.

We now find the number of atoms inside the volume bounded by the diameter of the lithium bubble, $d_s = 16$ μ, and by the length of the stopping path of lithium ions in a deuterium-tritium target, $l_{fr} = 30$ μ. The volume occupied by this tube segment is $V_b = \pi r_s^2 l_{fr} = 6 * 10^{-9}$ cm^3 and contains $N_{(D, T)} = 3 * 10^{14}$ atoms.

Next, we find the total energy introduced by the bubble (without the energy released by nuclear reactions, as well as taking into account that only half of the energy released goes inside): $W_{total} = (½) * 7$ MeV $* 2.25 * 10^{13} = 7.8 * 10^{19}$ eV. Dividing this energy by the total number of atoms in the volume bounded by the cylinder diameter $d_s = 16$ μ and by the length equal to the depth of cylinder penetration into the deuterium-tritium mixture $l_{fr} = 30$ μ, ($N_{(D, T)} = 3 * 10^{14}$), we determine that for each atom of the mixture there is approximately 260 keV energy.

The optimal energy near which the reaction yield is maximal equals 107 keV, the corresponding cross section being 5 barns, [1], p. 899.

Let the reaction yield Y be equal to the value $Y = \sigma n l_t$, where σ is the reaction cross section; n is the density of the target atoms; l_t is the mean path of the incident nucleus in the target. The yield is the probability of a nuclear reaction per one incident nucleus. We take the cross section of the reaction $D + T = He + n$ equal to 2 barns. Then, for the probability of the reaction Y to be equal to $Y = 1$, it is required that the product $n l_t$ be equal to $n l_t = 5 * 10^{23}$.

In our case the density of the target atoms $n_{(D, T)} = 5 * 10^{22}$, and the yield probability is unity at the interaction length $l_t = 10$ cm. It can be seen that the interaction length in this case is $l_t = 10$ μ, that is, four orders are lacking for the reaction yield to be 100%.

If the interaction of a lithium tube with the (D, T) target results in a radial compression of the target by 10^4 times, then the density will rise by 8 orders of magnitude; the interaction length will reduce by 4 orders; and the product nl_t will reach the value $nl_t = 5 * 10^{23}$.

The speed of the mixture deuterons at the energy $W_D = 260$ keV per one deuteron is: $\beta_D = V_D / c = (2W_D/Mc^2)^{1/2}$. For the deuteron: $\beta_D = 2.3 * 10^{-2}$ and $V_D = 7 * 10^8$ cm / s. The transverse dimension of the mixture after the compression by 10^4 times is $l_t = 10^{-7}$ cm, so the lifetime of plasma in a compressed state τ can be considered equal to: $\tau = l_t / V_D = 1.4 * 10^{-16}$ s.

The initial density of the mixture $n_{in} = 5 * 10^{22}$. After a radial compression by 10^4 times the density will increase up to $5 * 10^{30}$ atoms/cm^3, and the product $n\tau$ will be equal to $n\tau = 7 * 10^{14}$ atoms * s/cm^3, so the Lawson criterion is performed in this case. It should be kept in mind that we have assumed the energy per one atom to be equal to $W_D = 260$ keV, instead of 10 keV for which the Lawson criterion is formulated and where the reaction cross section is a thousand times smaller than at the maximum.

6. Conclusion

The total energy release in the (D, T) reaction is 17.6 MeV.
A 100% interaction of deuterium and tritium nuclei in the target ensures the energy release $\Delta W = 17.6 * 10^6 * 1.5 * 10^{14} = 3.5*10^{21}$ eV or approximately 560 J. Assuming that one gram of TNT releases 4 kJ of energy, the energy of a microburst in this case will be equal to 140 mg of TNT.

When the accelerator is operated in the continuous mode, $f_{0in} = 2.4 * 10^7$ tube segments can be accelerated per one second. If each segment releases 560 J of energy during the interaction with the deuterium-tritium target, the total energy release will be greater than 13 GJ / s or 13 GW.

The acceleration of the cone might provide best results for the compression of the (D, T) target in comparison with the compression of the tube segment. The cone needs to be accelerated with its base forward.

References

1. Tables of Physical Data. The Handbook, edited by I.K. Kikoin, Moscow, Atomizdat, 1976
2. S.N. Dolya, K.A. Reshetnikova, About the Electrodynamic Acceleration of Macroscopic Particles, JINR Communication P9-2009-110, Dubna, 2009, http://www1.jinr.ru/Preprints/2009/110(P9-2009-110).pdf
http://arxiv.org/ftp/arxiv/papers/0908/0908.0795.pdf
3. I.M. Kapchinsky, Particle Dynamics in Linear Resonance Accelerators, Moscow, Atomizdat, 1966

On measuring the size of nuclei of comets

Possibilities of measuring the size of nuclei of comets hidden by dust clouds are discussed. To this end, the dust cloud should be irradiated with a flow of rods accelerated in a linear mass accelerator to the velocity $V_{fin} = 6$ km / s. Each rod should be equipped with a transmitter with a power of 1 μW, which is destroyed in a collision with a comet's nucleus, or continues to work if the rod passes through the dust cloud without collision. Radio signals are received by three independent ground stations. At a distance of $R = 1000$ km from the nucleus of the comet the power of the received signals is $P_{sig} = 10^{-17}$ W, the receiver noise power: $P_{noise} = 10^{-20}$ W.

Introduction

To measure the size of nuclei of comets is very important since a comet, even flying at a considerable distance from the Earth, can cause serious damage.

Let us consider in more detail the possibility of exposing the dust cloud that hides the nucleus of the comet to the flow of rods, each equipped with a radio transmitter.

1. Acceleration of rods

1. 1. Parameters of the accelerated rod

We will consider acceleration of macroparticles shaped as a rod with a conical head, which are electrically charged.

Acceleration of rods in a helical waveguide is well studied [1]. This acceleration requires that the initial velocity of the rod and the phase velocity of the wave were approximately identical. As the acceleration of rods goes on, the wave phase velocity in the spiral waveguide should be increased so that rod was always in the same phase of the wave, called a synchronous phase. The phase velocity of the wave in the waveguide can be increased by increasing the coil pitch or decreasing the coil radius, or doing both at the same time.

Let the diameter of the rod be d_{sh} = 2 mm and length l_{sh} = 300 mm. Then the cross-section area of the rod is $S_{tr} = \pi \, d_{sh}^2 / 4 = 3.14 * 10^{-2}$ cm², and the volume of the rod is $V_{sh} \approx 1$ cm³. The mass of the rod at an average density ρ_{aver} = 5 g/cm³ is m_{sh} = 5 g.

1. 2. The ratio Z / A

We assume that the average atomic mass of the rod is A_{sh} = 30. The number of nucleons in the rod can be found from the proportion

$$6 * 10^{23} \text{ -------- } 30 \text{ g}$$
$$x \text{ ---------- } 5 \text{ g}$$

where x = 10^{23} atoms or $A_{sh} = 3 * 10^{24}$ nucleons.

We take the surface tension of the electric field on the rod to be $E_{surf}= 3 * 10^8$ V / cm. Using the formula for the surface tension of the field for the cylinder

$$E_{surf}= 2\kappa / r, \tag{1}$$

we find the charge density per unit length of the rod

$$\kappa = E_{surf} * r/2e = (5 * 10^7 * 0.1) / (5 * 10^{-10} * 300 * 2) = 10^{14}, \tag{2}$$

from which we can obtain

$$N_p = (\kappa / e) * l_{sh} = 3 * 10^{15}. \tag{3}$$

Thus, if $N_p = 3*10^{15}$ protons are set on the rod, the surface tension of the field will turn out to be $E_{surf}= 3 *10^8$ V / cm.

Now that we know the total number of excess protons on the rod $N_p = 3 * 10^{15}$ and the number of nucleons in it $A_{sh} = 3 * 10^{24}$ we can find the charge-to-mass ratio for the rod $Z / A = N_p / A = 3 * 10^{15} / 3 * 10^{24} = 10^{-9}$.

1.3. Proton beam irradiation of rods

To accelerate a rod shaped as a cylindrical rod with a pointed-cone head in a spiral waveguide, it should be electrically charged. The electric charge can be imparted to a rod by irradiating it with a beam of protons so that the irradiating protons remained on it. Then the electric charge of rods will increase in proportion to the proton beam current and the duration of exposure. Let the proton beam current be $I_{beam} = 0.5$ A, and the current pulse duration be $\tau_{beam} = 1$ ms. Then the total number of protons in the current pulse is $N_p = I_{beam} * \tau_{beam} / e = 3 * 10^{15}$ protons.

1.4. Proton beam irradiation of rods. Proton energy

Let a cylindrical rod gas-dynamically accelerated to $V_{in} = 1$ km / s is exposed to a proton beam from an external source. The field surface strength is assumed to be $E_{surf} = 300$ MV / cm. Then, for the diameter of the cylinder $d_{sh} = 2$ mm, we find that the minimum energy of the protons that can overcome the Coulomb repulsion of the protons previously placed on the rod should be $W_p > e E_{surf} * d_{sh} / 2 = 30$ MeV.

1. 5 . Proton beam irradiation. Mean free path

Protons with an energy of 30 MeV have a range in aluminum about 1 g/cm^2 [2, p. 953]. Given the density of aluminum $\rho_{Al} = 2.7$ g/cm^3, we find that the range of the protons in aluminum is $l_{Al} \approx 3$ mm. Since the average density of the material we have chosen for the cylinder is $\rho_{aver} = 5$ g/cm^3, about twice the density of aluminum, the mean free path of protons with an energy of 30 MeV in the rod will be approximately 2 mm.

Apparently, it is necessary to gradually increase the energy of the protons during the irradiation. As more and more protons are set on the rod, the energy of the protons emitted later, on the one hand, should be sufficiently high to overcome the Coulomb repulsion of the protons that are already on the rod and, on the other hand, should be such that the path of the protons in the material of the rod was much smaller than its diameter.

In this energy range the path of the protons in the material increases linearly with energy; for example, protons with energy $W_p = 3$ MeV, has a path $3 * 10^{-2}$

g/cm² [2, p. 953], or about 100 μ, and will not be able to cross the rod diameter of 2 mm. They will lose their energy by ionization of the material and would stay within the rods.

2. Acceleration length

The acceleration rate of a charge in an electric field can be written as

$$\Delta W = (Z / A) \, eE_{zw}, \qquad (4)$$

and for the strength of the wave $E_{zw} = 70$ kV / cm, the rate of the energy gain will be $\Delta W = 7 * 10^{-4}$ eV / (m * nucleon), so that the required increase in energy $\Delta \varepsilon = 0.2$ eV / nucleon will be attained over the length

$$L_{acc} \approx \Delta \varepsilon / \Delta W = 30 \text{ m}. \qquad (5)$$

3. Selection of the spiral waveguide parameters

The spiral waveguide is a standard coaxial cable with its central wire wound into a spiral. In such a cable, there is no dispersion in a wide range of frequencies, i.e., the velocity of propagation does not depend on the frequency and the phase velocity in this cable coincides with the group velocity.

The wave (pulse) propagation velocity V in this cable is determined by the tightness of winding of the central conductor into a spiral and the dielectric properties of the medium that fills the cable. This relation is called the dispersion equation

$$\beta = \text{tg } \Psi / \varepsilon^{1/2}, \qquad (6)$$

where $\beta = V / c$, V is the velocity of the pulse through the cable, $c = 3 * 10^{10}$ cm / s is the speed of propagation of electromagnetic waves in a vacuum, tg $\Psi = h/2\pi r_0$, h is the winding pitch of the spiral, r_0 is the radius of the spiral winding, ε it the relative dielectric constant of the medium filling the cable. The wave as is runs along the spiral in a circle of $2\pi r_0$, while moving a small distance h along the axis of the spiral. The wave further slows down due to the dielectric properties of the medium determined by the value of ε.

To accelerate a body by a pulse running in a cable, the insulator must be removed from inside the spiral, and then the pulse velocity in this cable will slightly increase [1] ,

$$\beta = \sqrt{2} * tg \, \Psi / \varepsilon^{1/2}. \qquad (7)$$

The pulse running through a line with distributed parameters contains not only the gradient of the magnetic field, which are accelerated magnetic dipoles, but also the electric field E_{zw}, which can accelerate a charged body.

The initial velocity of the projectile in a spiral $\beta_{sh\,in}$ expressed in terms of the speed of light $\beta_{in} = V_{in} / c$, where $c = 3 * 10^{10}$ cm / s is the speed of light in a vacuum, is $\beta_{in} = 3.3 * 10^{-6}$, and the final velocity is $\beta_{fin} = 2 * 10^{-5}$. Spiral should apparently consist of several sections, so that an optimal acceleration rate could be selected within each section. The acceleration wavelength can be determined from the condition $x = 2\pi r_0 / (\beta * \lambda_0) = 1$, where x is a dimensionless parameter in the arguments of the modified Bessel functions, r_0 is the radius of the spiral, β is the phase velocity , λ_0 is the acceleration wavelength in a vacuum , and $\lambda_0 = s/f_0$, f_0 is the acceleration frequency.

With the initial radius of the spiral equal $r_{0\,in} = 20$ cm, and the dielectric constant of the medium between the spiral and the screen $\varepsilon = 1280$, we find: $\lambda_0 = 3.8 * 10^7$ cm, $f_0 = 790$ Hz. Thus, the slow wavelength for the beginning of the acceleration is $\lambda_{slow} = \beta\lambda_0 = 1.25$ m.

3.1. Parameters of the spiral

In order to obtain the required field intensity E_0 in the spiral waveguide, we need a power to be introduced into it, which is defined by the formula [1]

$$P = (c / 8) * E_0^2 * r_0^2 * \beta * \{\}, \qquad (8)$$

where P is the high-frequency power introduced into the coiled waveguide, r_0 is the radius of the spiral, and β is the phase velocity of the wave , which is determined from the dispersion equation. The braces in (8) are

$$\{\} = \{ (1 + I_0K_1/I_1K_0) (I_1^2 - I_0I_2) + \varepsilon (I_0/K_0)^2 (1 + I_1K_0/I_0K_1) (K_0K_2 - K_1^2)\}, \quad (9)$$

where I_0, I_1, I_2 are the modified Bessel functions of the first kind , K_0, K_1, K_2 are the modified Bessel functions of the second kind . The first term in the braces corresponds to the flow propagating inside the spiral, and the second term corresponds to the flux traveling outside the spiral. Since the space between the spiral and the screen is filled with a dielectric, a factor ε appears in front of the second term [1].

In this case, deceleration of the electromagnetic wave to velocities of the order of the velocity of sound requires the use of both the geometrical properties of the structure (small-pitch spiral) and the properties of the medium, for which we chose the relative permittivity $\varepsilon = 1280$.

Thus, the flow of the high-frequency power propagating outside the spiral is more than 10^3 times higher than the power that propagates inside the spiral. Therefore, the first term inside the braces can be neglected, and the value of the braces for the argument $x = 1$ is approximately $\{\} \approx 4\varepsilon$.

In accelerators the synchronous phase is selected on the front slope of the pulse, so that the electric field accelerating the rod is always lower than the peak value. Let us choose the synchronous phase $\varphi_s = 45^0$, $\sin \varphi_s = 0.7$, $E_{zw} = E_0 \sin \varphi_s$. Thus, the amplitude of the wave which accelerates the cylindrical rod should be

$$E_0 = E_{zw} / \sin \varphi_s = 100 \text{ kV} / \text{cm}. \qquad (10)$$

Then, the wave power in watts expressed by the formula (8) is
$P (W) = 3 * 10^{10} * 10^{10} * 10^2 * 4 * 3.3 * 10^{-6} * 1.28 * 10^3 * 4 /$
$/(8 * 9 * 10^4 * 10^7) = 300 \text{ MW}. \qquad (11)$

3.2. Transition from a sine wave to a single pulse

This power is achievable for pulse technology. We expand the sinusoidal pulse [1], corresponding to the half-wave
$E_{pulse} = E_{0pulse} \sin (2\pi t / T_0)$, $2\pi / T_0 = \omega_0$, $\omega_0 = 2\pi f_0$ in a Fourier series.

$$f_1 (\omega) = (2 / \pi)^{1/2} \int_0^{T_0 / 2} \sin\omega_0 t * \sin\omega t \, dt. \qquad (12)$$

The pulse spectrum is narrow and covers the frequency range from 0 to $2\omega_0$. Since the spiral waveguide dispersion (dependence of the phase velocity on the frequency) is weak, it can be expected that the full range of frequencies from 0 to $2\omega_0$ will propagate at an approximately identical phase velocity. As a result, the half-wave sinusoidal pulse will spread out several-fold in space due to only an increase in the phase velocity of the wave. In this case the spiral waveguide should be matched with the supply feeder in the band $\Delta f \approx \omega_0 / 2\pi$.

We introduce the concept of the pulse amplitude $\tilde{U}$ connected with the field strength in the axis of the spiral E_0 by the relation [1],

$$\tilde{U}_{pulse} = E_{0pulse}\lambda_{slow}/2\pi, \quad \lambda_{slow} = \beta\lambda_0, \quad \lambda_0 = c/f_0. \qquad (13)$$

The wavelength choice $\lambda_0 = 3.8 * 10^7$ cm means, that we choose the duration of the acceleration of the rod ($f_0 = c/\lambda_0 = 790$ Hz), $\tau_{pulse} = 1 / (2f_0) = 630$ μs. The voltage pulse amplitude will be $\tilde{U} = E_0\lambda_{slow} / 2\pi = 2$ MV, and the pulse current through the coil windings will be $\tilde{I} = P / \tilde{U} = 150$ A. Table 1 summarizes the main parameters of the accelerator.

Table 1. Parameters of the accelerator

Z/A= 10^{-9}, dielectric outside spiral, wave power, P	P = 300 MW
	$\mu = 1$, $\varepsilon = 1280$
Velocity, initial – final, β_{ph}	$\beta_{ph} = 3.3*10^{-6} - 2*10^{-5}$
Initial radius of spiral, r_0	$r_0 = 20$ cm
Wave frequency, f_0,	$f_0 = 790$ Hz
Electric field strength, E_0	$E_0 = 100$ kV/cm
Accelerator length, L_{acc}	$L_{acc} = 30$ m
Pulse duration, τ	$\tau = 630$ μs
Voltage amplitude, $\tilde{U}_a$	$\tilde{U}_a = 2$ MV

3. 3. The capture of rods in the acceleration mode. Admission

Let us calculate the required accuracy to match the initial phase of the accelerating wave (pulse) with the synchronous phase. The theory of rod capture into a traveling wave gives [3], $\Delta\varphi = 3\varphi_s$, $(+ \varphi_s - 2\varphi_s)$. In practice, it means, for example, that in our case, where T / 4 corresponds to the duration of 316 μs or 90^0 degrees, one phase corresponds to the time interval of approximately 3 μs.

In linear accelerators the buncher gives bunches with the phase width $\pm 15^0$. To avoid large phase fluctuations, we require that the accuracy of synchronization of rods with the accelerating pulse be $\Delta\tau = \pm 15 * 3$ μs=

= ± 45 μs. This synchronization accuracy seems quite attainable for the gun powder start, that is, preliminary gas-dynamic acceleration of rods.

Let us now calculate the accuracy tolerance for coincidence of the initial rod velocity and the phase velocity of the pulse propagating along the helical structure. We introduce the quantity $g = (p-p_s) / p_s$, which is the relative difference of the pulses [3]. In the non-relativistic case is simply a relative velocity dispersion of $g = (V-V_s) / V_s$. The vertical scale of the separatrix is calculated by the formula [3]

$$g_{max} = ± 2 [(W_\lambda ctg\varphi_s/2\pi\beta_s) * (1 - \varphi_s / ctg\varphi_s)]^{1/2}, \qquad (14)$$

where $\varphi_s = 45^0 = \pi / 4$, $ctg\varphi_s = 1 [1 - \varphi_s / ctg\varphi_s]^{1/2} = 0.46 \ 2 * 0.46 = 0.9$, and $W_\lambda = (Z / A) eE_0\lambda_0 sin\varphi_s/Mc^2$.

Let us find $W_\lambda = (Z / A) eE_0\lambda_0 sin\varphi_s/Mc^2$, the relative energy gain of the rod over the wavelength λ_0 in vacuum. In our case $\lambda_0 = c/f_0 =$ =$3.8 * 10^7$ cm, $sin\varphi_s = 0.7$, $Mc^2 = 1$ GeV, and $W_\lambda = 2.66 * 10^{-5}$. Substituting numerical values, we get $g = (V_{in}-V_s) / V_s = \Delta V / V_s$, and, finally, $\Delta V / V_s = ± [2.66 * 10^{-5} / (6.28 * 3.3 * 10^{-6})]^{1/2} * 0.9 = ± 0.3$.

Thus, the tolerable discrepancy between the initial rod velocity and the pulse velocity is of the order of $\Delta V / V_s = ± 30\%$. For the initial rod velocity $V_{in} = 1$ km / s of the tolerable velocity deviation is $\Delta V < 300$ m / s.

4. Radial motion

As is well known [3], in azimuth - symmetric wave of phase stability, the phase stability region corresponds to the radial defocusing. In this case, we cannot have simultaneously radial and phase stability; under the phase stability conditions external field are required for radial focusing. In this phase regions the radial component of the electric field of the wave is directed towards the increasing radius, i.e., radially accelerates the rods.

In this region of rod velocities, hypersonic, hundreds of thousands of times lower than the speed of light, focusing by magnetic quadrupole lenses is not effective, and electrostatic quadrupole lenses are most suitable for this purpose. They focus the rods in one plane and defocus them in another. Collected into a doublet, two such lenses give the resulting focusing effect. The accelerator should be divided into separate sections, and the focusing doublets can be placed between the acceleration sections.

5. Carrying rods away beyond the Earth's atmosphere

5. 1. Lift

When the length of the accelerator is $L_{acc} \approx 30$ m, it should be placed horizontally. To carry a rod beyond the atmosphere, we can use a small asymmetry in the rod's shape, such that it creates a lifting force F_y. The relevant equation of vertical motion can be written as

$$mdV_y / dt = C_y \rho_0 V_x^2 * S_{tr} / 2, \qquad (15)$$

where C_y is the aerodynamic lift coefficient , $\rho_0 = 1.3 * 10^{-3}$ g/cm^3 is the air density at the surface of the Earth, V_x is the horizontal velocity of the rod, and S_{tr} is the cross-section of the rod.

5. 2. Ballistics. Air resistance

We calculate the motion of an electrodynamically accelerated rod. Equations of motion of the rod can be written as

$$mdV / dt = - \rho C_x S_{tr} V^2 / 2, \qquad (16)$$

where m is the mass of the rods , Vis the velocity , $\rho = \rho_0 e^{-z}/H_0$ is the barometric formula for changes in atmospheric density with height ,
$\rho_0 = 1.3 * 10^{-3}$ g/cm^3 is the air density at the surface of the Earth, $H_0 = 7$ km is the height in which the density decreases by e times.

The aerodynamic coefficient or the coefficient of drag is a dimensionless quantity that takes into account the "quality" form of rods,

$$C_x = F_x / (½) \rho_0 V_0^2 S_{tr}. \qquad (17)$$

Equation (16) can be written as

$$V(t) = V_0 / [1 + \rho C_x V_0 * S_{tr} * t/2m]. \qquad (18)$$

To calculate the rate of change of magnetic dipoles with time, it is necessary to find the aerodynamic coefficient C_x.

5. 3. The calculation of the drag coefficient for air rods

We assume that the rod is shaped as a cylindrical rod with a conical head. The impact of a nitrogen molecule at the sharp cone causes a change in the longitudinal velocity of the molecules

$$\Delta V_x = V_x * \Theta_h^2 / 2, \tag{19}$$

where Θ_h is the angle at the vertex of the cone. The molecules of the nitrogen to pass rod pulse:

$$p = mV = \rho V_x S_{tr} t * \Delta V_x. \tag{20}$$

The change in momentum per unit time - the power, the power of a frontal inhibition,

$$F_{x1} = (\tfrac{1}{2}) \rho V_x S_{tr} * V_x * \Theta_h^2. \tag{21}$$

Dividing F_{x1} by $(\tfrac{1}{2}) \rho V^2_x S_{tr}$, we get the drag coefficient for a sharp cone in the mirror image molecules from the cone (the Newton)

$$C_{x\,air} = \Theta_h^2. \tag{22}$$

Let the length of the conical part of the rod be $l_{cone} = 12.5$ mm and the diameter be $d_{sh} = 2$ mm. This means that the angle at the vertex of the cone is $\Theta_t = d_{sh} / l_{cone} = 1.6 * 10^{-1}$ and $C_{x\,air} = 2.5 * 10^{-2}$.

In order to have a pointed-cone head, the rod should be sufficiently long. Limiting the length of the rods is the fact that for a good efficiency of the acceleration length l_{tot} rods should be less than a quarter wavelength delayed $\lambda_{slow} = \beta\lambda_0$, i.e.: $l_{tot} < \beta\lambda_0 / 4$. In this case, for the begin of the acceleration, $\beta\lambda_0 / 4 = 40$ cm.

5. 4. Rod passage through the atmosphere

Let us draw up a table to show the time dependence of the vertical velocity, lift, and horizontal velocity of the rod. The vertical velocity is calculated by the formula

$$\Delta V_y = C_y \rho V_x^2 * S_{tr} * \Delta t / 2m. \tag{23}$$

The climb is calculated by the formula

$$H_{fly\ n+1} = H_{fly\ n} + V\tilde{\ }_y * \Delta t + C_y \rho V_x^2 * S_{tr} * (\Delta t)^2/4m, \quad (24)$$

where $V\tilde{\ }_y$ is the average vertical velocity in the time interval Δt. A decrease in the horizontal velocity over time will be described by formula

$$V_{x\ n+1} = V_{x\ n} / [\ 1 + (C_x \rho V_{xn} * S_{tr} * \Delta t/2m)]. \quad (25)$$

The change in the air density with altitude will be taken into account by the barometric formula $\rho = \rho_0 * \exp[-y/H_0]$, where $H_0 = 7$ km.

Table 2 shows the cylinder flight parameters as a function of time. The second column shows the vertical velocity of the cylinder, and the third shows the horizontal velocity of the cylinder, the fourth shows the height it gained after the corresponding second of flight, and the fifth shows the density of the atmosphere at this height.

Table 2. Flight parameters at $C_x, C_y = 2.5 * 10^{-2}$.

t, s	V_x, km/s	V_y, km /s	Y, km	ρ air, g/cm^3
0	6	0	0	$1.3*10^{-3}$
10	3.72	3.67	18	$2*10^{-4}$

In this case the time of gaining the maximum height is $\tau_{max} = V_y / g = 367$ s, where $g = 10^{-2}$ km/s^2 is the gravity acceleration, the flight range is $S = V_x * 2\tau_{max} = 2700$ km, and the maximum height is $Y = V_y^2/2g = 670$ km.

6. Rod flight path control

To control the flight path of the rod at its side surface is applied four quadrants of a material with different resonant absorption of laser radiation. "Right - left" and "up - down" deviations of the rod are effected by evaporating a corresponding quandrant exposed to laser radiation with a resonance wavelength.

In silicon with different degrees of doping the Langmuir frequencies ω_{pl} will be different and, therefore, these four quadrants will have different resonant frequencies of absorption [4].

Thus, we can assure that only one of the four quadrants, namely, that whose position is opposite to the direction from which the rod must be diverted, will evaporate upon resonant absorption of laser radiation and produce a jet.

In order to evaporate the quadrant in a short time, such that the heat from absorption of the laser radiation did not penetrate the body, the laser pulse must be sufficiently short.

6. 1. Parameters of the flight and trajectory change

Let the comet move as far from the surface of the Earth as $h_{com} = 200 - 400$ km and the distance from the comet to the floating platform carrying the accelerator and the laser is $s_2 = 1000$ km. Rod has a mass $m_b = 5$ g and moves with a velocity $V_b = 3$ km / s. When a mass $m_{jet} = 15$ mg moves at a velocity $V_{jet} = 1$ km / s perpendicular to the rod velocity, the transverse momentum transfer will be $p^{\perp} = m_{jet} * V_{jet}$, and this will result in the deflection angle $\theta^{\perp} = p^{\perp} / (m_b * V_b) = 10^{-3}$. This angle, on a distance from the target: $s_1 = 100$ km, will result in rejection of the body the trajectory: $\Delta l = s_1 * \theta^{\perp} = 100$ m.

6. 2 . Energy relations for the jet efflux

The silicon heat capacity is $c_{SI} = 20$ J / (mol * degree) [5, p. 199], melting point $T_{mel} = 1415$ ^{0}C, solid - liquid phase transition heat $\Delta H_m = 50$ kJ / mol, boiling point $T_{eva} = 3300$ ^{0}C, and liquid - vapor phase transition heat $\Delta H_m = 356$ kJ / mol [5 , p. 289]. Considering all energy needed for evaporation and the fact that 1 mole of silicon is 28 g, we find that evaporation of 1 gram of silicon requires an energy of ~ 15 kJ / g.

For the average directed velocity of silicon atoms to be $V_{jet} = 1$ km / s, the thermal velocity should be $V_T = 2.5$ km / s. Indeed, after averaging the velocity in one of the transverse plane, we get $V^{\sim}_1 = (V_T / \pi) * \int \sin\varphi \, d\varphi =$ $=(2 / \pi) * V_T$, where integration over angles is from 0 to π. After averaging in two transverse planes we obtain $V^{\sim}_2 = V_{jet} = (2 / \pi)^2 * V_T \approx 0.4 V_T$, so that , in addition to evaporation of silicon, it is necessary to impart thermal velocity $V_T = 2.5$ km / s to its atoms for their average directed velocity to be $V_{jet} = 1$ km / s.

We find the energy of the silicon atom moving with a velocity V_T from the relation $m_{Si} * V_T^2 / 2 = \varepsilon_{Si} = 1.5 * 10^{-19}$ J. Given that 1 g contains

$2 * 10^{22}$ atoms, we find that additional energy of the order of 3 kJ / g is required, and the total energy consumption should be $W_{las} \sim 20$ kJ / g.

6. 3. Irradiation parameters

Consider the abilities of an infrared laser pulse at a distance $s_2 = 1000$ km from the irradiated body. For the diffraction divergence of the laser beam to be sufficiently small, it is necessary to integrate individual laser emitters into a laser array [6], similar to phase locking of individual emitters in a phased-array antenna.

Let the total diameter of the laser array be $d_{gr} = 3$m. Then the diffraction divergence angle will be $\theta_{dif} = \lambda / d_{gr} = 3 * 10^{-6}$, where $\lambda = 10$ μ is the wavelength of the laser radiation. Thus, at the distance $s_2 = 1000$ km the laser spot area can be estimated as $S = \pi * (s_2 * \theta_{dif})^2 = 30$ m^2.

Let the area of the quadrant on the rod to be evaporated for diverting the rod to the angle $\theta^\perp = p^\perp / (m_b * V_b) == 10^{-3}$ be $s_s = 3$ cm^2. In this case the geometric factor is 10^{-5}. So, the laser energy required for heating and evaporating one gram of silicon is $W_{las1} = 20$ kJ. Considering the geometrical factor, the energy should be 10^5 times higher, $W_{las2} = 2 * 10^9$ J, and radiation energy required to heat and evaporate 15 mg of silicon will be $W_{las3} = 30$ MJ.

Thus laser irradiation of a rod at a distance $s_2 = 1000$ km from the laser followed by evaporation of one of four quadrants on the rod will impart a transverse momentum $p^\perp = m_{jet} * V_{jet}$ to the rod, which will give rise to the deviation angle $\theta^\perp = p^\perp / (m_b * V_b) == 10^{-3}$. At a distance $s_1 = 100$ km from the nucleus of the comet this angle will result in diversion of the rod from the unperturbed trajectory by $\Delta l = s_1 * \theta^\perp = 100$ m.

7. Rod penetration depth into the nucleus of a comet

Let us find the rod penetration depth into the nucleus of the comet from the following considerations. First, we calculate the depth of penetration into aluminum, and then extend the results to the nucleus of the comet.

We take the density of aluminum to be $\rho_{Al} = 2.7$ g/cm^3 [5, p. 99]; the mass of 1 g-mol of aluminum is 27 g. The specific heat of aluminum is

$c_{Al} = 24.35$ J / (mol * K) [5, p. 199], the melting point is $T_{mel} = 660$ C^0, the solid - liquid phase transition heat is $\Delta H_{mel} = 10.8$ kJ / mol [5, p. 289], and the boiling point is $T_{eva} = 2520$ ^{0}C. To bring aluminum to a boil required 50 kJ / mol, the liquid - vapor phase transition is heat $\Delta H_{eva} = 293$ kJ / mol, [5, p. 289].

Adding all up, we find that evaporation of one mole of aluminum requires 375 kJ, and evaporation of one cubic centimeter of aluminum requires 37.5 kJ/cm³. The cross sectional area of the rod is $S_{tr} = \pi d_{sh}^2 / 4 = 3.14 * 10^{-2}$ cm²; thus, 1.2 kJ are needed for evaporation of an aluminum cylinder with a diameter equal to the diameter of the rod and a length of 1 cm.

The kinetic energy of the rod $mV_m^2 / 2$ at the velocity $V_m = 3$ km / s is $E_{kin} = 2.25 * 10^4$ J. From the data on reactions of meteorites with solids [7] it is known that at the velocity about 3 km / s, meteorites spend about 20% of their kinetic energy for evaporation of substance. In our case it means that is consumed by evaporation and 4.5 kJ will be spent for evaporation and the rod penetrates aluminum as deep as $l_{pen} = 4.5$ kJ / (1.2 kJ / cm) ≈ 3 cm.

The nucleus of the comet has a lower density than aluminum but on hitting it, the rod will certainly be destroyed, at least the transmitter will stop emitting electromagnetic energy.

It will thus be possible to determine whether the rod passed through the dust cloud surrounding the nucleus of the comet, or collided with the nucleus of the comet, in which case the rod will be destroyed and stop transmitting radio signals.

8. Radio emission of rods

Suppose that the transmitter attached to the rod emits radio waves at a wavelength $\lambda_{rad} = 3$ cm, (half -wave dipole) with a power $P_{rad} = 1$ μW. The parameters of the radiating pulse are taken to be the following: the pulse width $\tau_{pul} = 10$ μs and the repetition rate $F_{rep} = 100$ Hz. At a distance of $R = 1000$ km within the area of the receiving antenna $S_{ant} = 100$ m² the power of the received signal will be $P_{res} = P_{rad} * S_{ant}/4\pi R^2 = 10^{-17}$ W.

The noise power of the receiver, if the receiver is a superconducting cavity with $T_{cav} = 1$ K^0, will be $P_{noise} = kT_{cav}\Delta f$, where

$k = 1.38 * 10^{-23}$ J / degree is the Boltzmann constant, $T_{cav} = 1$ K^0, $\Delta f_1 = 1 / \Delta \tau_{pul} = 10^5$ Hz is the reception band. Since every second there will come $F_{rep} = 100$ Hz, the reception band can be taken to be two orders of magnitude narrower, $\Delta f_2 = 10^3$ Hz. This band corresponds to the resonator Q-factor $Q = f_{rad} / \Delta f_2 = 10^7$, where $f_{rad} = c / \lambda_{rad} = 10^{10}$ Hz is the frequency of the radiation. As a result, the received signal will exceed the energy noise by 3 orders of magnitude.

The Q-factor of modern superconducting resonators is of the order of $Q = 5 * 10^{10}$ [8]. This means that the intrinsic bandwidth of the resonator is a few hundredths of a hertz, $\Delta f_{cav} = f_0 / Q$, and it will have to be specially extended to allow appropriately fast processing of the received signals.

9. Operation of the system equipment

Figure 1 shows a diagram of the equipment.

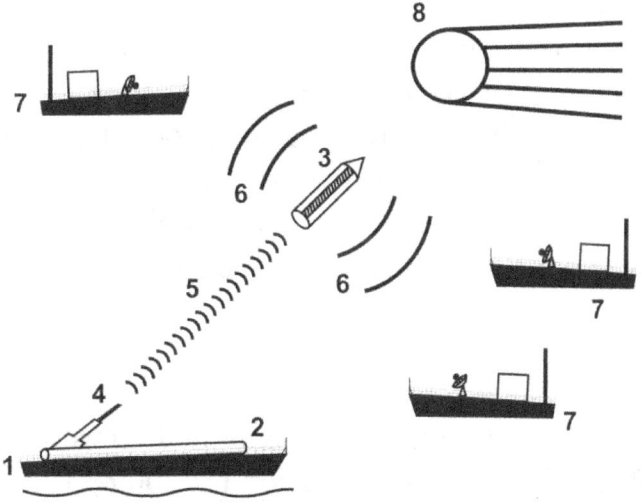

Fig.1. (1)- floating platform, (2)- accelerator, (3)- rods, (4)- four - channel IR laser, (5)- laser beam, (6)- radio waves emitted by a rod,
(7)- three independent stations receiving the emitted radio waves,
(8)- comet.

The equipment operates as follows.

The electromagnetic mass accelerator (2) are located on the floating platform (1). It accelerates rods (3) to the velocity $V_{in} = 6$ km / s. Due to the pointed cone head and small asymmetry rods penetrate the Earth's atmosphere. The rods

101

having an azimuthally and vertical speed approximately equal: $V_{\varphi, r} \approx 3.7$ km / s. Rods carry four quadrants of silicon with different degrees of doping, so that each quadrant has its own resonant frequency of absorption. To control the flight path of a rod, it is irradiated (5) with four-channel infrared laser (4). Rods radiate radio waves (6) with the wavelength $\lambda_{rad} = 3$ cm and power $P_{rad} = 1$ μW. Radio waves are received by the receivers' with sensitivity 10^{-20} W, arranged on three floating platforms (7), determining the spatial coordinates of the rods. Rods pass through the dust cloud that hides the nucleus of the comet (8) without breaking. To meet with the nucleus of the comet, rods destroyed and cease to emit radio waves.

Conclusions

Receiving signals by three independent receivers can allow us to restore accurately the spatial coordinate of the emitting rods, and the presence or absence of the signal will allow us to judge whether the rod hit the nucleus of the comet or not. If rod, after passing through the dust cloud that hides the comet nucleus, continues to emit radio signals, it is precisely the area of the dust cloud. If after passing through the dust cloud the rod ceased to emit radio signals, it collided with the nucleus of the comet. Measuring the spatial coordinates of the rods that left the dust cloud hiding the nucleus of the comet and knowing the coordinates of the stopped rods, we can calculate the size of the comet nucleus.

References
1. S. N. Dolya, K. A. Reshetnikova, About the Electrodynamic Acceleration of Macroscopic Rods, JINR Communication P9-2009-110, Dubna, 2009,
http://www1.jinr.ru/Preprints/2009/110(P9-2009-110).pdf
http://arxiv.org/ftp/arxiv/papers/0908/0908.0795.pdf
2. Tables of Physical Data. The Handbook, edited by I.K. Kikoin, Moscow, Atomizdat, 1976
3. I.M. Kapchinsky, Rod Dynamics in Linear Resonance Accelerators, Moscow, Atomizdat, 1966
4. V.Y. Timosheko, Nano-optics, Lecture 4, Exciton and impurity absorption of light, Moscow State University, Research and Education Center for Nanotechnology,
http://nano.msu.ru/files/systems/V/autumn2011/optics/Timoshenko_L04_NOC 2011.pdf

5. Tables of Physical Data, The Handbook, edited by I. S. Grigor'ev
and E. Z. Meylikhov, Moscow, Nuclear Power Published, 1991
6. V. P. Kandidov, Laser grid,
http://www.pereplet.ru/nauka/Soros/pdf/9912_068.pdf
7. L. S. Novikov, Exposure to particulate matter of natural and artificial origin
on the spacecraft, University Book, Moscow, 2009, p. 58,
http://window.edu.ru/library/pdf2txt/658/74658/54484/page8
8. http://www.linearcollider.org/about/Publications/Reference-Design-Report,
http://en.wikipedia.org/wiki/Superconducting_radio_frequency

Artificial Micrometeorites

An iron ball, a beryllium sphere and a tungsten tube segment with diameter $d = 20$ μ, are electrically charged while proton beam irradiating. These bodies are accelerated by the running pulse field in a spiral waveguide up to velocity: $V_m = 30$ km / s. The accelerator, generating micrometeorites is placed at satellites on the Earth orbit. This article considers processes of penetration of micrometeorites into the Earth atmosphere. It is shown that micrometeorites evaporate at the height of 100 - 150 km from the surface of the Earth. A micrometeorite which is a segment of the beryllium tube equipped with a graphite cone in the head part is the very meteorite to reach the Earth surface without being broken.

Introduction

Artificial micrometeorites, which are electrically charged and accelerated by electrodynamics method, may be applied to study fluctuations in the atmospheric density and distribution of the magnetic and electric fields in it. We will consider micrometeorites, consisting of various elements and having a different shape, but they will all have the same diameter $d_m = 20$ μ and the same velocity: $V_m = 30$ km / s.

We assume that the accelerator generating micrometeorites is located on a satellite rotating over the near-by orbit of the Earth.

1. Acceleration of iron balls

We consider how to accelerate the iron ball having diameter $d_{Fe} = 20$ μ up to velocity $V_m = 30$ km / s, or if the velocity is expressed in the units of the velocity of light, $\beta_{fin} = 10^{-4}$, where $\beta = V / c$, $c = 3 * 10^5$ km / s, the velocity of light in vacuum.

To speed up this ball by the field of the electromagnetic wave, it must be electrically charged.

It is possible to charge the iron ball electrically by electrons bombarding it, which should stay on the ball thus transferring the electric charge to it. The field emission will limit the amount of electrons on the ball in the process. Starting from the certain field strength due to the field emission the set of electrons will leak out from the ball no matter how many electrons and new electrons we would plant, they will all leave the ball.

The value of the surface threshold for the current density of iron is: $e\varphi = 4.3$ eV [1], page 444, and the electric field strength $E_{surf} = 30$ MV / cm in this case is $j = 10^{-1}$ A/cm^2 [1], page 461. If you cover the iron ball with platinum and passivate it with oxygen, the surface threshold will increase up to $e\varphi = 6.5$ eV [1], page 445, and the leakage current density (for values $E_{surf} = 30$ MV /cm will reduce to: $j = 10^{-9}$ A/cm^2 [1], page 461.

However, with increasing field strength up to 40 MV / cm the leakage current density will increase to $j = 10^{-4}$ A/cm^2, and at higher growth of the surface field strength up to value $E_{surf} = 50$ MV / cm, the current density will increase to the value of $j = 10^{-1}$ A/cm^2 [1], page 461. So that, in achieving this field strength, all new electrons planted on the ball will leak out due to the field emission.

It is evident that it is impossible to achieve the high field strength by placing a large electric charge on the ball, when it is irradiated by the electron beam. This is due to the fact that the electron is a light particle and it easily leaves the ball which has a large surface electric field.

The solution is as follows: it is necessary to irradiate a ball with heavy particles such as protons. We will proceed from the surface electric field $E_{surf} = 300$ MV / cm. It should be borne in mind that the proton is a heavy particle and while irradiating the ball with protons the ball will be transferred a large transverse (relative to the direction of acceleration) pulse. Therefore, irradiation of the ball must be performed by two opposite beams of protons to exclude the transverse pulse.

1. 1. The number of protons placed on the ball

We find the number of protons, placed on the ball, on the Coulomb law:

$$E_{surf} = N_p \, e/r_{Fe}^2, \qquad (1)$$

where E_{surf}= 300 MV / cm is the surface field strength on the ball,
e = 4.8 * 10^{-10} - elementary charge, expressed in CGS units,
r_{Fe} = 10 μ - ball radius, N_p - the number of protons, placed on the ball.

Substituting numbers into the formula (1) we find that it is required to place N_p = 2 * 10^9 protons to obtain the surface field strength E_{surf} =300MV/cm.

For the protons would overcome the electrostatic repulsion of the protons previously placed on the ball, the proton energy must be higher than:

$$W_p > E_{surf} * r_{Fe} = 300 \text{ keV}. \qquad (2)$$

The running length of the protons with the energy of 300 keV in iron can be estimated as R_{Fe} <1 μ [1], page 953, which is less than the radius of the ball. If the running length is greater than the radius of the ball, the proton energy should be increased gradually having been slowed down by the Coulomb field of the ball and the protons would remain on the ball.

1. 2. Number of nucleons in the ball

In the accelerator technology an important parameter is the ratio of charge to the mass of the accelerated particle where the charge is expressed in units of the elementary charge (electron, proton), and the mass is expressed in the units of the mass of the nucleon.

We find the mass of the iron ball. The volume of the ball is: $V_{Fe} = (4/3) \, \pi \, r^3_{Fe}$, $V \approx 4 * 10^{-9}$ cm^3, the density of iron is assumed to be: ρ_{Fe}= 8 g/cm^3, so that the mass of the ball is $m_{Fe} \approx 3 * 10^{-8}$ g. The number of nucleons in the ball is found from Avogadro's law:

$$6*10^{23} \text{ ------------ } 56 \text{ g}$$

$$x \text{ ------------}3*10^{-8} \, g, \qquad (3)$$

where: 56 g - mass of one gram mole of iron. From (3) we find that
x = 6 * 10^{23} * 3 * 10^{-8}/56 atoms or A = 6 * 10^{23} * 3 * $10^{-8} \approx 2 * 10^{16}$ nucleons.

The ratio of the charge to the mass (Z / A) for the ball is as follows:

$$eN_p/A = Z/A = 2*10^9/2*10^{16} = 10^{-7}. \qquad (4)$$

1. 3. Electrostatic acceleration of the balls

Usually linear accelerators of heavy charged particles are constructed as follows. First, preliminary electrostatic acceleration is performed at a certain speed, till which the electromagnetic wave can be slowed down and further the final acceleration of the traveling wave is performed. In our case, since the ratio of the charge to the mass of the ball is extremely small (10^{-7}) it is needed to uses a structure where the ultra slow electromagnetic waves can propagate. Recall that the singly charged ion of Uranium - 238 has a ratio of the charge to the mass $1/238 = 4.2 * 10^{-3}$ that is much bigger than in our case. Let us select the voltage for the platform on where the store with the balls is located, to be equal to $U_{el.\,st.} = 100$ kV.

The balls from the store should be produced and mechanically accelerated till the velocity of about one meter per second. Thereafter, during the motion they must be irradiated by two opposite proton beams so that the number of protons placed on the ball should be equal to $N_p = 2 * 10^9$. After that, the balls must be accelerated by electrostatic potential difference $U_{el.\,st.} = 100$ kV.

Let the field strength of the iron ball be $E_{el.\,st.} = 10$ kV / cm. The equation of motion of the ball with a specific (per nucleon) charge $Z / A = 10^{-7}$, can be written as follows:

$$dV/dt = (Z/A)eE_{el.\,st.}/M, \qquad (5)$$

where M is the mass of the nucleon. Assuming the initial velocity of the ball is equal to zero, we obtain an expression for the dimensionless velocity $\beta = V / c$, where $c = 3 \times 10^5$ km / s - the velocity of light in vacuum. We write the dependence of the dimensionless velocity of time:

$$\beta = \{(Z/A)*eE_{el.\,st.}/Mc^2\}*ct. \qquad (6)$$

The final (electrostatic acceleration) velocity can be found from the following relation:

$$\beta^2 = 2(Z/A)e\,U_{el.\,st}/Mc^2, \qquad (7)$$

where we find: $\beta^2 = 2*10^{-11}$, $\beta = 4.5*10^{-6}$.

1. 4. Electromagnetic acceleration of iron balls

We choose a spiral waveguide as the slow-wave structure, which can propagate ultra slow electromagnetic waves [2].

1. 4. 1. The required delay of the electromagnetic wave

To speed up charged particles by using the electrodynamics way, it is necessary that the initial velocity of the particles and waves coincide approximately: $\beta = \beta_{ph}$, and it is required to increase the phase velocity of the wave while particle acceleration.

The slowdown of the waves in the tight helix is approximately equal to the ratio of the perimeter to the pitch of the helix. The dispersion equation relating the phase velocity with the size of the spiral β_{ph} in the tight helix looks like this:

$$\beta_{ph} = tg\Psi , \qquad (8)$$

where $\beta_{ph} = V_{ph} / c$, $tg\Psi = h/2\pi r_0$ - tangent of the winding angle, h - the winding pitch of the helix, r_0 - the radius of the spiral.

It is also known that in the medium where the permittivity and permeability have significant values, the phase velocity of the electromagnetic wave is less than the velocity of light in vacuum and equal to the following:

$$\beta_{ph} = 1/(\varepsilon\mu)^{1/2}. \qquad (9)$$

We can expect that if you put the spiral into this medium, the general slowdown of the electrostatic wave will increase and the electromagnetic wave phase velocity will be equal to:

$$\beta_{ph} = tg\Psi /(\varepsilon\mu)^{1/2}, \qquad (10)$$

where factor $tg\Psi$ determines the slowing properties of the structure and $(\varepsilon\mu)^{1/2}$ - shows what properties of slowdown the medium possesses. It is evident that considerable quantities $(\varepsilon\mu)^{1/2}$, according to formula (10) may be prepared by slowing by the order of 10^5.

1. 4. 2. Exact solutions

Now we consider the properties of helical waveguide, fully immersed to the medium with permittivity ε and permeability μ. From Maxwell's equations for the internal components of the field area we find the following field components for the inner area of the spiral:

$$E_{z1} = E_0 I_0(k_1 r)$$
$$E_{r1} = i(k_3/k_1)E_0 I_1(k_1 r)$$
$$H_{\varphi 1} = i\varepsilon(k/k_1)E_0 I_1(k_1 r)$$
$$H_{z1} = -i(k_1/\mu k)tg\Psi I_0(k_1 r_0)E_0 I_0(k_1 r)/I_1(k_1 r_0)$$
$$E_{\varphi 1} = -tg\Psi I_0(k_1 r_0)E_0 I_1(k_1 r)/I_1(k_1 r_0)$$
$$H_{r1} = (k_3/\mu k)tg\Psi I_0(k_1 r_0)E_0 I_1(k_1 r)/I_1(k_1 r_0). \tag{11}$$

The field components for the helix external area can be written in the form [2]:

$$E_{z2} = I_0(k_1 r_0)E_0 K_0(k_1 r)/K_0(k_1 r_0)$$
$$E_{r2} = -i(k_3/k_1)I_0(k_1 r_0)E_0 K_1(k_1 r)/K_0(k_1 r_0)$$
$$H_{\varphi 2} = -i(k/k_1)I_0(k_1 r_0)E_0 K_1(k_1 r)/K_0(k_1 r_0)$$
$$H_{z2} = i(k_1/k)tg\Psi I_0(k_1 r_0)E_0 K_0(k_1 r)/K_1(k_1 r_0)$$
$$E_{\varphi 2} = -tg\Psi I_0(k_1 r_0)E_0 K_1(k_1 r)/K_1(k_1 r_0)$$
$$H_{r2} = (k_3/\mu k)tg\Psi I_0(k_1 r_0)E_0 K_1(k_1 r)/K_1(k_1 r_0), \tag{12}$$

where the omitted factor is $e^{i(\omega t - k_3 z)}$.

The dispersion equation, the relationship between the phase velocity of the wave, the structure parameters and the wave frequency for the helix, fully immersed in Ferro dielectric medium is as follows:

$$ctg^2\Psi = k_1^2/k^2\{I_0(k_1 r_0)K_0(k_2 r_0)/I_1(k_1 r_0)K_1(k_2 r_0)\}, \tag{13}$$

where: $k_1 = k(1/\beta_{ph}^2 - 1)^{1/2}$, $k_2 = k(1/\beta_{ph}^2 - \varepsilon\mu)^{1/2}$, $k = \omega/[c(\varepsilon\mu)^{1/2}]$. For the large slow down, the dispersion equation is simplified and looks like this:

$$\beta_{ph} = tg\Psi/(\varepsilon\mu)^{1/2}. \tag{14}$$

This formula, as well as for the helix located in vacuum has a simple physical meaning: the ratio of the phase velocity to the velocity of the wave in the medium is the ratio of the lengths to the wave running along the helix and its axis. After simple calculations we obtain the relation between the power flux in the spiral and E_0 field tension on its axis in the case of the spiral emerged into medium c with permittivity ε [2]:

$$P = (c/8)\, E_0^2\, r_0^2\, [\, kk_3/k_1^2\,]\, \varepsilon\{(1+I_0K_1/I_1K_0)(I_1^2-I_0I_2) +$$

$$+ (I_0/K_0)^2(1+I_1K_0/I_0K_1)(K_0K_2-K_1^2)\}. \qquad (15)$$

This relationship between the flux and the electric field strength on the axis coincides with the expression for the spiral in vacuum except for factor ε which determines the dielectric constant of the medium.

1. 4. 3. Partial filling of the medium

If the insulator is only outside the helix region and inside the helix is free of the filling, formula (15),relating the power flux with the electric field strength on the axis, is somewhat different [2]:

$$P = (c/8)\, E_0^2\, r_0^2[\, kk_3/k_1^2\,]\{(1+I_0K_1/I_1K_0)(I_1^2-I_0I_2) +$$

$$+ \varepsilon\, (I_0/K_0)^2(1+I_1K_0/I_0K_1)(K_0K_2-K_1^2)\}. \qquad (16)$$

Only the second term in the curly brackets, the corresponding power flux propagating beyond the spiral is multiplied by ε. The factor ε before the bracket does not contain it. We call this case - partial filling of the spiral with the medium.

The dispersion equation relating the phase velocity of the wave with frequency is as follows if the moderating medium is beyond the helix [2]:

$$ctg^2\Psi = (k_1k_2/k^2)\varepsilon\mu\{I_0(k_1r_0)K_0(k_2r_0)/I_1(k_1r_0)K_1(k_2r_0)\}F_0, \qquad (17)$$

where:

$$F_0 = \varepsilon\{1+(k_1/k_2)\mu\, I_0K_1/I_1K_0\}*[1+(k_1/k_2)\varepsilon\, I_0K_1/I_1K_0]^{-1}. \qquad (18)$$

Arguments of the functions $I_{0,\,1}$ are: (k_1r_0), functions $K_{0,\,1}$ are: (k_2r_0).
The most interesting case for us is tight winding, thus, the dispersion equation for this case is simplified and looks as follows:

$$\beta_{ph} = \text{tg}\Psi \, F_0^{1/2}/(\varepsilon\mu)^{1/2}. \qquad (19)$$

It is easy to see that in the case of ε, $\mu = 1$, the dispersion equation becomes the vacuum case: $\beta_{ph} = \text{tg}\Psi$. In the general case the $F_0^{1/2} > 1$ and retarding properties in the spiral filled Ferro dielectrics outside are worse than in the case with full immersion.

In the important case of ε, $\mu \gg 1$ the expression for F_0 is simplified:

$$F_0 = \mu, \qquad (20)$$

and the dispersion equation becomes:
$$\beta_{ph} = \text{tg}\Psi/\varepsilon^{1/2}. \qquad (21)$$

In the most interesting case is when large dielectric ε, $\varepsilon \gg 1$ is outside the helix and the medium is not ferromagnetic $\mu = 1$, the expression is simplified for $F_0 = 2$, and the dispersion equation looks like this:

$$\beta_{ph} = \sqrt{2} \, \text{tg}\Psi/\varepsilon^{1/2}. \qquad (22)$$

This very case providing the large slowdown and high electric field E_0 on the axis will be considered below in detail.

Maintaining synchronism between the wave velocity and particle velocity can be achieved by increasing the helix pitch. For a spiral wound on the cylindrical surface it would lead to thin winding and that is why the electric field strength (for a fixed power generator) will reduce. There is another variant of making the spiral waveguide. It is possible to wind the spiral onto the narrowing cone. In fact in this case there is an increase of the phase velocity achieved not by increasing the helical pitch h, but by reducing radius r_0. At the same factor (kk_3/k_1^2) r_0^2, equal to βr_0^2, remains approximately constant.

1. 4. 4. Acceleration length

Finite velocity $V_m = 30$ km / s corresponds to the velocity β_{fin}, expressed in units of the velocity of light, $c = 3 * 10^5$ km / s, which is equal to: $\beta_{fin} = 10^{-4}$. The energy per nucleon in this case is: $W_{fin} = Mc^2 \beta_{fin}^2 / 2 = 5$ eV / nucleon, where $Mc^2 = 1$ GeV, the nucleon mass, expressed in energy units. When acceleration rate $eE_0\sin\varphi_s = 250$ keV / cm, where E_0 – the peak value of the field strength, $E_0 = 300$ kV / cm, $\varphi_s = 60^0$ synchronous phase, $\sin\varphi_s = 0.87$, this energy can be achieved over the length: $L_{acc\,Fe} = W_{fin} / [(Z/A)\,eE_0\sin\varphi_s] = 2$ m.

1. 4. 5. Power consumption

The relationship between the power flux and the electric field strength for a spiral when the space between the spiral and the screen filled with a dielectric and the screen is given by (16).

The argument of the modified Bessel functions in curly brackets in (16) is the value of $x = 2\pi r_0/\lambda_{slow}$, where r_0 - radius of the helix, $\lambda_{slow} = \beta\lambda_0$, β - phase velocity in a spiral, $\lambda_0 = s/f_0$ - wavelength in free space, $f_0 = \omega/2\pi$ - wave frequency, ω - angular frequency of the wave.

From formula (16) we see that in order to have the maximum field strength E_0 for the fixed power flux P and at fixed r_0 and β, we must have the smallest value of the curly bracket which is simply a numerical factor in the formula (16).

During acceleration there is the rate change β of the ball and it leads to the fact that for the given power the electric field strength decreases. To maintain a uniform rate of acceleration within the same section (and as will be seen later, the whole accelerator should be divided into separate sections), it is necessary to reduce the radius of the spiral winding along the length of the section [2]. To reach this, it is needed to wind the spiral onto the narrowing cone so that the product βr_0^2 remains approximately constant over the length of the section.

Then, at a reasonable length of the section, parameter $x = 2\pi r_0/\beta\lambda_0$ rapidly decreases This is due to the fact that not only the velocity of the ball β increases but also the radius of the spiral winding r_0 reduces. Table 1 shows the values of the first (I) and second (II) terms in the curly brackets of equation (16).

Table 1.

x	I	II
0.1	0.1	66.8
0.2	0.14	22
0.3	0.18	12.14
0.4	0.226	8.286
0.5	0.273	6.365
0.6	0.326	5.277
0.7	0.386	4.618
0.8	0.454	4.208

0.9	0.532	3.958
1	0.620	3.819
1.1	0.721	3.763
1.2	0.836	3.774
1.3	0.968	3.844
1.4	1.119	3.96
1.5	1.29	4.142
1.6	1.494	4.369
1.7	1.724	4.650
1.8	1.989	4.989
1.9	2.295	5.393
2	2.69	5.867
2.5	5.441	9.68
3	11.336	17.601

Recall that the second term in the curly brackets corresponds to the power flux propagating between the spiral and the screen. This volume should be filled with a dielectric having a large dielectric constant ε, so that the second term in the curly brackets would be much larger than the first one. Acceleration can begin within one section at a value of $x = 2$, and complete at $x = 0.5$.

The maximum field strength will then be obtained for a value of $x = 1$, that is in the middle of the section. The field strength at the beginning and end of section will be less than [2]. Everything discussed above, applied to the coil without an external screen. The exact formulas that take into account the effect of the screen, obtained in [3]. Clear, however, that if the screen is far enough away from the spiral $R_{screen} > 3\ r_0$, then its effect on the propagation of radio waves in the spiral slightly.

We choose the parameter x to $x = 2$ and calculate the power required to create the field $E_0 = 300\ kV\ /\ cm$ for the beginning of the acceleration, i.e. $\beta_{in} = 4.5 * 10^{-6}$. We choose to determine the initial spiral radius $r_0 = 1$ cm. Value $x = 2$ means that we chose $2\pi r_0/\beta_{in}\lambda_0 = 2$, $\lambda_{slow} = \pi r_0 = 3.14$ cm, then we have chosen to start accelerating the vacuum wavelength $\lambda_0 = \pi r_0/\beta_{in} = 7 * 10^5$ cm. This wavelength corresponds to a frequency $f_0 = c\ /\ \lambda_0 = 43$ kHz.

As we have already noted, besides the purely geometric spiral wave slowdown the wave should be additionally slowed down further. To do this, the space between the spiral and the outer shield must be filled with dielectric medium having relative permittivity ε. For barium titanate, near the Curie point, the achievable values of $\varepsilon = 8 * 10^3$, [4], page 557. We take value ε slightly smaller, namely: $\varepsilon = 1.28 * 10^3$.

Substitute the numbers into the formula (16).

$P\ (W) = 3*10^{10}*3*10^5*3*10^5*4.5*10^{-6}*1.28*10^3*5.87/8*3*10^2*3*10^2*10^7 =$
$= 12.6 *10^6$ W.

There is a large number of high-frequency generators having this electric power, obtained in the previous expression. However, if to accelerate a single bulb, it can be accelerated by the field travelling pulse along the spiral waveguide. Duration of the base of the sinusoidal pulse corresponding to frequency f_0, is: $\tau_{pulse} = 1/2f_0 \approx 10$ μs. The pulse amplitude can be found from the following relation:

$$U_{pulse} = E_0 * \lambda_{slow}/2\pi = 314 \text{ kV}. \tag{23}$$

1. 4. 6. The spiral pitch

Since we have chosen the spiral radius equal to: $r_0 = 1$ cm, then it is required to take a very small spiral winding pitch to get the slowdown equal to $\beta_{ph\ in} = 4.5 * 10^{-6}$ in this spiral, where $\beta_{ph\ in} = V_{in} / c$ - initial phase velocity, expressed in terms of the velocity of light coinciding with the initial velocity of the ball.

According to formula (22) for the start of the spiral where the velocity of the balls is $\beta_{ph\ in} = 4.5 * 10^{-6}$, $r_0 = 1$ cm, $\varepsilon = 1.28 * 10^3$, replacing tg $\Psi \approx h/2\pi r_0$, where h - the helix winding pitch, we find,

$$4.5*10^{-6} = \sqrt{2}*h/(2\pi r_0 * \varepsilon^{1/2}).$$

From this expression we define that the helix winding pitch should be equal to:

$$h = 7\ \mu. \tag{24}$$

The amplitude of the electric field strength E_0, which has been chosen, is: $E_0 = 300$ kV / cm and 30 V / μ. At step $h = 7$ μ. There is a danger of interterm dielectric breakdown since the interterm voltage is 30 V / μ * 7 μ = 210 V. The breakdown voltage of the polyimide is 300 MV / m [4], page 550, or 300 V / μ, so that we can choose the helix structure as follows: the copper coil with a cross section $\Delta = 6 \mu$ and polyimide insulation thickness of 1 micron.

1. 4. 7. Electric power damping capacity while pulse propagating in the spiral

The wave attenuation in the helical waveguide will lead to the fact that the amplitude of the pulse propagating in the spiral will decrease while pulse propagating from the beginning to the end of the spiral. This decrease is related with the Ohm current losses for heating the spiral.

I_φ current flows through the spiral up and, in fact, the Ohm losses are as follows:

$$\Delta P \text{ (W/turn)} = \frac{1}{2} I_\varphi^2 * R, \tag{25}$$

where: I_φ - current expressed in amperes, R is the loop resistance in Ohms. Then ΔP / coil - is expressed in Watts.

First, we find the electrical resistance of one coil. Resistance is calculated according to the conventional formula: $R = \rho l / s$, where $\rho = 1.7 * 10^{-6}$ Ohm * cm, the resistivity of copper: $l = 2\pi r_0$ - coil length, r_0 - radius of the helix, s - cross-section of the coil. Since the current flowing through the spiral is alternative (AC), then in formula (25) factor $\frac{1}{2}$ appears. The alternative current in the conductor penetrates to the depth of the skin - layer, which must be found.

The expression for this depth of the skin - the layer can be written as follows:

$$\delta = c/(\sqrt{2\pi\sigma\omega_0}), \tag{26}$$

where: $c = 3 \times 10^{10}$ cm / s - the velocity of light in vacuum , $\sigma = 5.4 * 10^{17}$ 1 / s - conductivity of copper , $\omega_0 = 2\pi f_0$ - circular frequency , $f_0 = 43$ kHz - frequency of the wave propagating in the spiral. Substitution of the numerical values in formula (26) gives: $\delta = 0.03$ cm.

The obtained skin depth - layer $\delta = 0.03$ cm, much larger than the distance between the coils of the spiral $h = 7 * 10^{-4}$ cm. This means that to reduce the resistance of one and, accordingly, to reduce attenuation of the coil, it is necessary to wind a rather wide tape with width $H = 2\delta = 0.06$ cm. Recall that we have chosen the ribbon thickness equal to $\Delta = 6 \mu$.

Then the resistance of one coil $R = \rho l / s$ will be equal to:

$$R = \rho * 2\pi r_0 / (2\delta * \Delta) = \rho * \pi r_0 / (\delta * \Delta). \qquad (27)$$

Substituting the numerical values for the start of the spiral, we find:

$$R = 1.7 * 10^{-6} * 3.14 / (3 * 6 * 10^{-6}) = 0.3 \quad \text{Ohm}. \qquad (28)$$

Now we find I_φ - current flowing through the coils. To do this, we use the following formula:

$$H_{zsurf} = (4\pi/c) n I_\varphi, \qquad (29)$$

where: H_{zsurf} - the magnetic field on the surface of the helix.

We find the relation between the component of the electric field E_0 on the helix axis and the component of the magnetic field $H_{z\,surf}$ on the spiral surface: $H_{z\,surf} = (k_1 / k) \, \text{tg}\Psi * I_0 (k_1 r_0) E_0 I_0 (k_1 r) / I_1 (k_1 r_0)$, [2]. For the inner area of the spiral, where k_1 - transverse wave vector: $k = (\omega / c) * \varepsilon^{1/2}$ - the wave vector, r_0 - the radius of the helix, the expression (k_1 / k) is equal to: $(k_1 / k) = 1/\beta_{ph}$, $\text{tg}\Psi \approx h/2\pi r_0$, so that $(k_1 / k) * \text{tg}\Psi = \varepsilon^{1/2}$. For $k_1 r_0 = 1$ the ratio $I_0 (k_1 r_0) / I_1 (k_1 r_0) = 2$, i.e. $H_{z\,surf} = 2 \varepsilon^{1/2} E_0$.

Thus, the component of the electric field: $E_0 = 300 \text{ kV / cm}$, on the helix axis corresponds to the magnetic field strength $H_{z\,surf} = 70 \text{ kGs}$ on the surface of the helix.

Now we can find the current flowing through the coils of the spiral. The $n I_\varphi$ current can be found from the relation: $n I_\varphi$ (A / cm) $= H_{zsurf} / (4\pi / c) = = (1.226)^{-1} * H_{zsurf} (\text{A / cm}) = H_{zsurf} (\text{Gs})$. Thus, the current in one coil is equal to:

$$I_\varphi (A) = H_{zsurf}(\text{Gs})/n, \qquad (30)$$

where: $n = 1 / h$ - the number of coils per one cm of the spiral.

In our case, $n = 1/7 \, \mu = 1.43 * 10^3$ coils / cm.

Substituting the numerical values in the formula (30), we find that the current in one coil is: $I_\varphi (A) = [70 \text{ kA / cm}] / (1.43 * 10^3 \text{ turns / cm}) \approx 50 \text{ A / turn}$.

115

The Ohm losses of the current in one coil are equal to the following:

$$\Delta P \text{ (W/turn)} = \tfrac{1}{2} I_\varphi^2 \ast R = 375 \text{ W/coil.} \tag{31}$$

Since there are n coils per 1 cm, the energy losses per 1 cm will be by n times more than in one coil:

$$\Delta P \text{ (W/cm)} = \tfrac{1}{2} I_\varphi^2 \ast R \ast n = 536 \text{ kW/cm.} \tag{32}$$

We introduce the ratio:

$$\Delta P/P = -2\alpha, \tag{33}$$

whence:

$$1/\alpha = L_{damping} = 2P/\Delta P = 2\ast 12.6 \ast 10^6 / 0.536 \ast 10^6 = 47 \text{ cm.} \tag{34}$$

This is the length over which the electric field strength is reduced by factor e due to attenuation. It can be seen that the motion of the balls when accelerating should be calculated taking into account the electric power. The accelerator itself should be divided into separate sections.

1. 4. 8. Radial motion of the balls

When particles are accelerated in an azimuthally symmetric field, it is not possible to achieve simultaneously both the radial and phase stabilities in the wave field [5]. The region of the phase stability corresponds to radial defocusing. The radial motion has the following form:

$$r (t) = r_{in} \exp [\gamma t], \tag{35}$$

i.e., any initial deviation from the acceleration axis r_{in} exponentially increases with time. The growth increment γ is equal to [5]:

$$\gamma = \pi f_0 \{ W_\lambda \operatorname{ctg} \varphi_s / \pi \beta_{ph} \}^{1/2}, \tag{36}$$

where: $W_\lambda = (Z / A) eE_0\lambda_0 \sin\varphi_s / Mc^2$ is the energy rate at a wavelength in vacuum. We substitute the numbers for the beginning of the acceleration and find:

$$W_\lambda = 10^{-7}\ast 3\ast 10^5 \ast 7\ast 10^5\ast 0.87/10^9 = 1.8\ast 10^{-5},$$

$$\gamma = 3.14*4.28*10^4\{1.8*10^{-5}*1.74/3.14*4.5*10^{-6}\}^{1/2} = 2*10^5.$$

The inverse value of γ corresponds to the rise time of the radial deflection by factor e, $1/\gamma = 5*10^{-6}$ s. If we recall that the initial velocity of the iron ball is $V_{in} = 1.35$ km / s, it is possible to find the spatial increment of the initial deviation from the following relation:

$$l_{defl} = (1/\gamma)*V_{in} = 0.675 \text{ cm.} \qquad (37)$$

We get the length of the initial rise of the spatial deviation to be much smaller than the acceleration length $L_{acc} = 2$ m, that means that in this case it is necessary to introduce radial focusing. Let us consider focusing of the accelerated iron balls by means of quadrupole lenses.

It is known [5], the quadrupole lens focusing the particles in one plane defocuses them in another one. If two lenses are turned relatively each other by 90^0, then in each of the transverse planes there are areas of the accelerator which alternately generate defocusing and focusing. Under certain conditions, such a system of lenses turns out to be the focusing.

Indeed, a particle moving accurately along the axis is not affected by any forces. The farther from the axis is the particle, the bigger values are the effective forces. Let the particle hit the first focusing section. Then it will bend the trajectory so to pass the defocusing section at a lower value of the field and the focusing forces will be stronger than the defocusing ones. A similar effect occurs in the case when the particle is first passing via the defocusing site. The resulting effect of the pair of the quadrupole lenses will be collecting, [5].

Focusing and defocusing action of the lenses is determined by their rigidity:

$$K = [(Z/A)eGl_1^2/Mc^2\beta_z], \qquad (38)$$

where (Z / A) is the ratio of the charge to the mass, G - gradient of the electric or magnetic field in the lens, l_1 – lens length, $Mc^2 = 1$ GeV - the rest mass of the nucleon, β_z - expressed in units of the velocity of light, the longitudinal velocity of the ball. Rigidity in this setting is unitless.

In contrast to the pair of quadrupole lenses, the accelerating section works as lens defocusing in the both perpendicular planes. Near the axis of acceleration there is no electric space charge and the following the condition:

$$\text{div } E = 0, \tag{39}$$

it shows the ratio between the longitudinal and transverse electric fields:

$$E_r = - (r/2)dE_z/dz, \tag{40}$$

This is evident, however, from the structure of the field in the spiral waveguide [2].

By analogy with the quadrupole lenses for the accelerating field the gradient is: $G_s = \frac{1}{2} dE_z / dz = \pi E_0/\lambda_{slow}$, where E_0 - the amplitude of the accelerating field, λ_{slow} - slow wavelength in the structure.

At first we consider focusing of particles by electrostatic quadrupole lenses. We require that the rigidity of the quadrupole lenses to have a greater rigidity of the accelerating section. This means that the particles in the deflection angle lens must be larger than in the deflection angle of the section. Indeed, the angle of deflection of the particle in the accelerating section is always directed outside the section. The section defocuses the accelerated particles in both transverse planes. The particle on the focusing plane article focusing on the plot should get the deflection angle inside to be at the same location on the focusing plane, she will receive an additional deflection angle outward. [5]

This means that the gradient field in the lens multiplied by the square of its length must be greater than the gradient field in the accelerating section, also multiplied by the square of its length:

$$G_l * l_l^2 > G_s * l_s^2. \tag{41}$$

The most difficult conditions for focusing occur at the beginning of the accelerator.

Let the lens length to be one-third of the length of the section, i.e. $l_s^2/l_l^2 = 10$, $l_s = 3l_l$. Then, the electric field gradient in the electrostatic lenses must at least be 10 times bigger than the electric field gradient in the sections, i.e:

$$G_l > 10 \, G_s. \tag{42}$$

118

Substituting the numbers to start the acceleration where $\lambda_{slow} = 3.14$ cm, we find that the electric field gradient in the electrostatic lenses must be greater than: $G_l > 3 * 10^6$ V/cm^2.

The ratio between the focusing by the electrostatic quadrupole lenses and magnetic quadrupole lenses can be represented as follows [5]:

$$G_e \text{ (V/cm}^2) = 300 \, \beta \, G_m(\text{Gs/cm}), \tag{43}$$

where this coefficient 300 corresponds to the transition from units in Gs units to the units in V / cm. Coefficient β is expressed in terms of the velocity of light, the longitudinal velocity of the particle. It is included into this formula as well as it enters the Lorentz force.

Substituting the value of $\beta = 4.5 * 10^{-6}$, we find that the magnetic field gradient in the quadrupole lenses must satisfy the following condition: $G_m > 2.2 * 10^9$ Gs / cm.

Let the length of the accelerating section be equal to: $l_s = 3$ cm. This length should be smaller than the attenuation length $L_{damping} = 47$ cm, which in this case is fulfilled. Then the length of the quadrupole lens is $l_l = 0.3 \, l_s = 1$ cm. Now we can calculate the stiffness of the lens $G_l * l_l^2$, expressed in more familiar units –i.e. Tesla multiplied by meter. In this case, the value is as follows: $G_l * l_l^2 = 22$ T * m.

Such high gradients in electrostatic and magnetic quadrupole lenses are "payment" for the high rate of acceleration. If such field gradients in the lenses are rather difficult to obtain, it is necessary to use a lower rate of acceleration. Placing of the focus elements increases the length of the accelerator by approximately twice.

2. Acceleration of the beryllium sphere

We have to say that from the point of view of accelerating the solid body acceleration is not optimal. This is due to the fact that for a solid body with the increasing radius of the ball a very important parameter Z / A. rapidly decreases. The charge on the bead is distributed on its surface, while the mass is concentrated in the ball volume and the ratio of the surface to the volume becomes smaller with increasing of the ball radius.

If from the acceleration of the solid ball to go to the acceleration of the sphere where all the mass is concentrated near the surface of the sphere, it can be expected that the ratio Z / A in this case is much larger than for the ball. We will consider the conditions of accelerating the beryllium sphere with radius $r_{Be} = 10$ μ, the same as for the iron ball. The equal thickness of the sphere: $\delta_{Be} = 1$ μ. is taken.

2 . 1. The number of protons placed on the sphere

We find the number of protons located on the sphere on the Coulomb law:

$$E_{surf} = N_p\, e/r_{Be}^2,$$

where $E_{surf} = 300$ MV / cm – the surface electric field strength on the sphere, e = 4.8 * 10⁻¹⁰ – is the elementary charge, expressed in CGS units, $r_{Be} = 10$ μ - radius of the sphere, N_p - the number of protons located on the sphere.

Substituting numbers into this formula, we find that to obtain the surface electric field strength $E_{surf} = 300$ MV / cm, it is required to have the following number of protons: $N_p = 2 * 10^9$. It is clear that for the same surface electric field strength and the same radius of the sphere and of the ball we get the same number of protons, which should be placed on the sphere or the ball.

2. 2. Number of nucleons contained in the sphere

We find the mass of the beryllium sphere. The volume of the sphere is equal to: $V_{Be} = 4\pi\, r_{Be}^2 * \delta_{Be} \approx 1.26 * 10^{-9}$ cm³, the density of beryllium is assumed to be $\rho_{Be} = 1.84$ g/cm³, so that the mass of the ball is $m_{Be} \approx 2.3 * 10^{-9}$ g. The number of nucleons in the sphere is found from Avogadro's law:

$$6*10^{23} \text{ ------------ } 9 \text{ g}$$

$$x \text{ ------------} 2.3*10^{-9} \text{ g,} \qquad (44)$$

where: 9 g - mass of one gram mole of beryllium. From (44) we find that x = 6 * 10²³ * 2.3 * 10⁻⁹/9 atoms or A = 6 * 10²³ * 2.3 * 10⁻⁹ =

$=1.4 * 10^{15}$ nucleons.

The ratio of the charge to the mass (Z / A) for the sphere is equal to the following:

$$eN_p/A = Z/A = 2*10^9/1.4*10^{15} = 1.4*10^{-6}. \qquad (45)$$

So we have obtained a the ratio of the charge to the mass to be by14 times greater than that for the iron ball, that means that for the same accelerator parameters it is 14 times shorter.

3. Acceleration of the tungsten tube segment

It is necessary to note that concerning the depth of penetration of substances into the body, the spherical shape of the body is not optimal. To increase the penetration depth, it is needed to use an elongated arrow-like -shaped body. Thus, it is possible to increase the mass of the body without increasing its cross-section.

Let us consider the conditions to accelerate a segment of the tungsten tube with a diameter $d_W = 20$ μ, wall thickness $\delta_W = 1$ μ and length $l_W = 1$ cm.

3. 1. The number of protons placed on the segment of the tube

We find the number of protons placed on the segment of the tube according to the Coulomb law:

$$E_{surf} = (2\kappa\, e/r_W), \qquad (46)$$

where $E_{surf} = 300$ MV / cm is the surface electric field strength on the segment of the tube, $e = 4.8 * 10^{-10}$ is the elementary charge, expressed in CGS units, $r_W = 10$ μ -the segment radius, κ - the number of protons per one centimeter length of the tube. Substituting the numbers into this formula in the CGS system:

$$10^6 = 2\, e\, \kappa/10^{-3},$$

we find that it is required to have the number of protons equal to: $\kappa = N_p = 10^{12}$ to reach the surface electric field strength $E_{surf} = 300$MV/cm. The tube section should be covered on the front and back parts with the tungsten hemispheres to have no sharp edges of the segment of the tube where large electric overvoltage could be expected.

3. 2. Number of nucleons in the tube segment

We find a mass of the tungsten tube segment. The volume of the segment is equal to: $V_W = 2\pi r_W * \delta_W * l_W \approx 6.28 * 10^{-7}$ cm³, the density of tungsten is assumed to be $\rho_W = 19$ g/cm³, so that the mass of the segment is: $m_W \approx 1.2 * 10^{-5}$ g. The number of nucleons in the tube segment is found from Avogadro's law:

$$6*10^{23} \text{------------} 184 \text{ g}$$

$$x \text{------------} 1.2*10^{-5} \text{ g,} \tag{47}$$

where 184 g - mass of one gram mole of tungsten. From (47) we find that $x = 6 * 10^{23} * 1.2 * 10^{-5}/184$ atoms or $A = 6 * 10^{23} * 1.2 * 10^{-5} = 7.2 * 10^{18}$ nucleons.

The ratio of the charge to the mass of the nucleon is (Z / A), thus the tube segment is equal to:

$$eN_p/A = Z/A = 10^{12}/7.2*10^{18} = 1.4*10^{-7}. \tag{48}$$

The obtained ratio of the charge to the mass is 1.4 times greater than for the iron ball that means that for the same parameters of the accelerator it will be by 1.4 times shorter than the accelerator for the iron balls, namely: $L_{acc\ W} = 1.4$ m.

4. Interaction of artificial micrometeorites with the magnetic and gravitational fields of the Earth

When moving in the Earth magnetic field the charged iron ball, like any charged particle, will be deviated by this magnetic field.
We take the magnetic path length approximately equal to the radius of the Earth $L_m = 7 * 10^3$ km, the magnetic field of the Earth is assumed to be $H_E = 0.5$ Gs. Then the deflection angle in the magnetic field and for this length of the iron ball will be equal to:

$$\Theta_m = (Z/A)e\ L_m* H_E/Mc^2 = 10^{-7}*7*10^8*0.5*3*10^2/10^9 \approx 10^{-5}. \tag{49}$$

122

It is clear that the magnetic field of the Earth does not influence the motion of the iron ball.

For the same length of the track $L_{grav} = L_m = 7 * 10^3$ km, we find the deflection angle of the iron ball by the gravitational field of the Earth is also very small:

$$\Theta_{grav} = g\, L_{grav}/V^2_m = 10^{-2}*7*10^3/30*30 = 7.7*10^{-2}. \qquad (50)$$

5. The depth of penetration of artificial micrometeorites in the Earth atmosphere

We assume that the main reason of destruction of artificial micrometeorite in the Earth atmosphere is their heating and evaporation due to this heating. It is assumed that the same amount of energy is needed for heating micrometeorite and atmospheric gas of the Earth.

5. 1. Penetration of the iron ball into the Earth atmosphere

We calculate the motion of the ball, taking into account the air resistance. The equation of the motion of the ball can be written in the following form (excluding the force of gravitational attraction to the Earth):

$$m dV/dt = - \rho C_x S_{tr} V^2/2, \qquad (51)$$

where: m – is the mass of the ball, V-velocity, $\rho = \rho_0 e^{-z/H0}$ - barometric formula changes in atmospheric density with altitude $\rho_0 = 1.3 * 10^{-3}$ g/cm³ - the air density on the Earth surface, $H_0 = 7$ km - the height at which the density decreases by factor e, C_x - drag coefficient. Let us assume that the value for the ball is: $C_x = 1$. In our case for the ball, $S_{tr} = \pi r^2_{Fe} = 3.14 * 10^{-6}$ cm². The solution of equation (51) can be written as follows:

$$V(t) = V_m/[1 + \rho V_m * S_{tr} * t/2m]. \qquad (52)$$

If to sum up the energy required for heating the ball from the room temperature to the melting point [4], page 289, the energy of the phase transition "solid – liquid", the energy that must be used for heating of the ball till its boiling point and, finally, the energy of the phase transition of the liquid to vapor [4], page 289, we will get the value of energy: $\Delta E_{Fe} = 7.9$ kJ / g.

The kinetic energy of the ball can be found from the following considerations: the energy per nucleon in the ball is 5 eV / nucleon, the ball contains $2 * 10^{16}$ nucleons, so that the total energy is equal to 10^{17} eV or 0.624 J.

Multiplying the energy that must be expended to evaporate one gram of iron per the weight of the iron ball $m_{Fe} = 3 * 10^{-8}$ g, we find that for the evaporation of the iron ball it is needed to spend:
$\Delta W_{evap} = 7.9$ kJ / g $* 3 * 10^{-8}$ g $= 2.4 * 10^{-4}$ J. This is a very small part of the kinetic energy of the ball: $W_{kin} = 0.624$ J.

The energy losses are related with the losses of the velocity by the following ratio: $\Delta E / E = 2 \Delta V / V$. It means that when the relative velocity decreases $\Delta V / V$ the amount $\Delta W_{evap} / W_{kin} = 2.4 * 10^{-4}/0.624 = 3.8 * 10^{-4}$, the iron balls evaporate. The magnitude of $\Delta V / V$ can be determined from the formula (52):

$$\Delta V/V = \rho V_m * S_{tr} * t/2m. \qquad (53)$$

After replacing $V_m dt$ by dz, we will integrate the value ∞ to z_0, where z_0 is the height, measured from the surface of the Earth, where the iron ball will evaporate. When integrating we obtain the following:

$$H_0 *\int_{\infty}^{(z0/H0)} \exp(-z/H_0) \, d(z/H_0) = H_0 \exp(-z_0/H_0). \qquad (54)$$

Thus, we have obtained an exponential equation to determine the value of z_0:

$$\rho_0 * S_{tr} * H_0 \exp(-z_0/H_0) /2m = \Delta W_{evap}/ W_{kin}, \qquad (55)$$

where: $\rho_0 = 1.3*10^{-3}$ g/cm^3, $S_{tr} = \pi r^2_{Fe} = 3.14*10^{-6}$ cm^2, $H_0 = 7$ km.

Substituting numbers into the equation (55) we obtain:

$$1.3*10^{-3}*3.14*10^{-6}*7*10^{5}* \exp(-z_0/H_0) /(2*3*10^{-8}) = 3.8*10^{-4},$$

From where we find:

$$\exp(-z_0/H_0) = 3.8*10^{-4}/5.15*10^4 \approx 10^{-8},$$

$$0.434*(-z_0/H_0) = -8, \quad z_0 = 130 \text{ km.}$$

Thus, at the height $z_{0\,Fe} = 130$ km from the Earth surface the iron ball evaporates.

5. 2. Penetration of the beryllium sphere into the Earth atmosphere

Summing up all the energy in a similar way as for beryllium, we find that the energy that must be expended to vaporize one gram of beryllium is equal to the following: $\Delta E_{Be} = 38$ kJ / g [5], page 289.

Taking into account the same reasons we find the kinetic energy of the sphere. The energy per nucleon in the sphere is: 5 eV / nucleon, the sphere contains $1.4 * 10^{15}$ nucleons, so that its kinetic energy is equal to $7 * 10^{15}$ eV or $W_{kin} = 4.37 * 10^{-2}$ J.

Multiplying the energy that must be expended to evaporate one gram of the beryllium sphere by the mass of the beryllium sphere $m_{Be} = 2.3 * 10^{-9}$ g, we find that for evaporation of the beryllium sphere the following energy must be expended: $\Delta W_{evap} = 38$ kJ / g $* 2.3 * 10^{-9}$ g $= 8.7 * 10^{-5}$ J. Substituting the numbers into equation (55), we find that the ratio $\Delta W_{evap} / W_{kin} = 2 * 10^{-3}$,

$$1.3*10^{-3}*3.14*10^{-6}*7*10^5*\exp(-z_0/H_0)/(2*2.3*10^{-9}) = 2*10^{-3},$$

$$6.2*10^5 \exp(-z_0/H_0) = 2*10^{-3}, \quad \exp(-z_0/H_0) = 3*10^{-9}, \quad 0.434*(z_0/H_0) \approx 9.$$

Where $z_0 = 150$ km, i.e., the beryllium sphere evaporates at height $z_{0\,Be} = 150$ km from the Earth surface.

5. 3. Penetration of the tungsten tube segment into the Earth atmosphere

We determine the kinetic energy of the tungsten tube segment and define the amount of energy in the electron – volts taking into account that the tube section contains $7.2 * 10^{18}$ nucleons, and the kinetic energy per nucleon is 5eV. Finally, we obtain $E_{kin} = 7.2 * 10^{18} * 5 = 3.6 * 10^{19}$ eV. Multiplying this value by $6.24 * 10^{-18}$, we obtain the value of the kinetic energy in Joules: $W_{kin} = 224$ J. The value of the energy that must be expended to evaporate one gram of tungsten is: [4], page 289, $E_{evap} = 5$ kJ / g, to vaporize the tungsten tube segment with mass $m_W = 1.2 * 10^{-5}$ g, it is needed to spend $W_{evap} = 6 * 10^{-2}$ J. This means that $W_{evap} / W_{kin} = 2.7 * 10^{-4}$ of the kinetic energy of the tube segment must be expended for evaporation. Substituting this value in equation (55), we obtain:

$$1.3*10^{-3}*3.14*10^{-6}*7*10^5*\exp(-z_0/H_0)/(2*1.2*10^{-5})=2.7*10^{-4},$$

$$\exp(-z_0/H_0)=10^{-6}, \quad 0.434*(z_0/H_0)=6, \quad z_0=100 \text{ km}.$$

This means that the length of the tungsten tube evaporates at height $z_{0\,W}=100$ km from the surface of the Earth.

5.4. Penetration of the tungsten tube in aluminum

We take the density of aluminum equal to $\rho_{Al}=2.7$ g/cm^3 [4], page 99, the mass of one gram moles of aluminum is 27 g. The specific heat of aluminum is, $c_{Al}=24.35$ J / (mol * K) [4], p.199 , the melting point $T_{mel}=660$ C^0, heat of phase transition solid - liquid is , $\Delta H_{mel}=10.8$ kJ / mol, [4] , page.289 , the boiling point is $T_{eva}=2520$ C^0. To reach the boiling point of aluminum, it is necessary to expend 50 kJ / mol, the heat for the phase transition "liquid – vapor" is $\Delta H_{eva}=293$ kJ / mol, [4], p. 289.

If you sum up all the required energy, we find that the evaporation of one mole of aluminum requires 375 kJ; the evaporation of one cubic centimeter of aluminum is 37.5 kJ/cm^3. The cross-section of the tungsten tube segment is $S_{tr\,W}=2\pi r_W * \delta_W=6.28 * 10^{-7}$ cm^2, so, to evaporate one centimeter of aluminum, it is required to spend: $W_{evap\,1\,cm\,Al}=2.35 * 10^{-2}$ J / cm. Recall that evaporation of the tungsten tube segment consumes $W_{evap\,W}=6 * 10^{-2}$ J. Assuming that the kinetic energy of the tube segment is consumed in equal proportions for evaporation of tungsten and aluminum, we find the depth of penetration of the segment in the aluminum, after which the tungsten tube segment evaporates: $l_{W\,evap}$,

$$l_{W\,evap}= W_{W\,evap} / W_{evap\,1\,cm\,Al}=6*10^{-2}/2.35*10^{-2}=2.5 \text{ cm}. \qquad (56)$$

It is seen that, in this case, the main reason limiting the depth of penetration of the body into the material is not the loss of kinetic energy of the body but its evaporation.

5.5. Penetration of the beryllium segment into the Earth atmosphere

We consider penetration into the atmosphere of the beryllium tube segment having the same dimensions as tungsten: the diameter of the tube $d_{Be} = 20$ μ, the thickness of the wall $\delta_{Be} = 1$ μ, the length $l_{Be} = 1$ cm. Recall that in front and at the end this segment of the tube should be covered by hemispheres to avoid sharp edges of the segment. The mass of such a segment is less than the mass of the tungsten segment in respect of specific weights of tungsten and beryllium and it is equal to: $m_{Be\ tube} = 1.15 * 10^{-6}$ g. The segment contains: $A_{Be\ tube} = 7 * 10^{17}$ nucleons.

We assume that the beryllium tube segment is electrically charged till the same surface electric field strength - $E_{surf} = 300$ MV / cm, for this purpose we have to place $N_p = 10^{12}$ protons. The ratio of the charge to the mass for this segment will be equal to $Z / A = 1.43 * 10^{-6}$.

For the beryllium tube segment the ratio of the charge to the mass is approximately the same as for the beryllium sphere.

The kinetic energy of the segment is: $E_{kin} = 7 * 10^{17} * 5 = 3.5 * 10^{18}$ eV. Transferring this value in Joules we obtain $W_{kin} = 3.5 * 6.24 = 22$ J.

To vaporize the beryllium tube segment, it is needed to consume $\Delta W_{evap} = 38$ kJ / g $* 1.15 * 10^{-6}$ g $= 4.37 * 10^{-2}$ J. The ratio of the energy required for evaporation to the kinetic energy of the tube segment is as follows:

$$\Delta W_{evap} / W_{kin} = 2 * 10^{-3}.$$

Substituting this value in the right-hand side of equation (55), we obtain:

$$1.3 * 10^{-3} * 3.14 * 10^{-6} * 7 * 10^5 * \exp(-z_0/H_0)/(2 * 1.15 * 10^{-6}) = 2 * 10^{-3},$$

$$\exp(-z_0/H_0) = 10^{-6},\ 0.434 * (z_0/H_0) = 6,\ z_0 = 100 \text{ km}.$$

The beryllium tube segment will evaporate at height $z_{0\ Be\ tube} = 100$ km, while the beryllium sphere evaporates at height $z_{0\ Be} = 150$ km, that shows the advantage (from the point of view of penetration into the material) of the elongated shape in comparison with the spherical body.

5.6. Penetration of the beryllium tube segment with a graphite cone in the head part

In [6] it is shown that using a sharp cone in the cylinder head results in significant reduction in the drag coefficient C_x. The connection between the cone angle, Θ head, and drag coefficient is given by [6]:

$$C_x = \Theta^2{}_{head}. \qquad\qquad (57)$$

Thus, for the values of the angle $\Theta_{head} = 10^{-3}$ the obtained drag coefficient $C_x = 10^{-6}$.

Into the head of the beryllium tube segment which is electrically insulated from it, we place a graphite cone with length $l_{cone} = 2$ cm. Let the diameter of the cone d_C coincide with the diameter of the beryllium tube segment $d_C = 20 \mu$, and let it be equal to the thickness of $\delta_C = 1 \mu$. The cone of this length is the very cone: $\Theta_{head} = d_C / l_{cone} = 10^{-3}$.

The lateral surface area of the cone is: $S_{cone} = (\frac{1}{2}) \pi d_C l_{cone} = 6.28 * 10^{-3}$ cm^2, volume $V_{cone} = 6.28 * 10^{-7}$ cm^3, mass $m_{cone} = 1.4 * 10^{-6}$ g, where we took the density of graphite to be equal to $\rho_{grap} = 2.25$ g/cm^3, [4], page 100. This cone contains $A_C = 6 * 10^{23} * 1.4 * 10^{-6} = 8.4 * 10^{17}$ nucleons. Then the total number of nucleons in the beryllium tube segment and graphite cone will be equal to: $A_{total} = (7 + 8.4) * 10^{17} = 1.54 * 10^{18}$ nucleons and the ratio of the charge to the mass is equal to $Z / A = 10^{12}/1.54 * 10^{18} = 6.5 * 10^{-7}$, that is approximately by 6.5 times greater than for the iron ball . This means that the accelerator for the segment of the beryllium tube with a sharp graphite cone at the head part of this segment is by 6.5 times shorter than the accelerator for the iron bead, whose length is equal to: $L_{acc\ Fe} = 2$ m.

Taking the specific heat of graphite $C_C = 8.54$ J / (mol * degree) [4], page 199 and the temperature at which graphite is destroyed, to be equal to: $T_{des\ C} = 3700$ C^0, we find that for the destruction of one gram of graphite it is needed to spend $W_{des\ C} = 8.54 * 3700/12 = 15.8$ kJ / g. To destroy the cone, it will be necessary to spend $W_{C\ cone} = W_{des\ C} * m_{cone} = 2.2 * 10^{-2}$ J.

The total energy of the system plus the beryllium tube segment with a cone is as follows: $A_{total} * 5$ eV $= 1.54 * 10^{18} * 5$ eV $= 7.7 * 10^{18}$ eV or $W_{kin} = 7.7 * 6.24 = 48$ J. Thus, the ratio of the energy required to destroy the cone, to the total kinetic energy of the system is equal to: $W_{C\ cone} / W_{kin} = 2.2 * 10^{-2}/48 = 4.6 * 10^{-4}$. The total mass of the system is: $m_{total} = 2.55 * 10^{-6}$ g.

Now we substitute all the numbers in the equation (55), where additional factor $C_x = 10^{-6}$ appears:

$$10^{-6}*1.3*10^{-3}*3.14*10^{-6}*7*10^5*\exp(-z_0/H_0)/(2*2.55*10^{-6}) = 4.6*10^{-4},$$

$$\exp(-z_0/H_0) \approx 1, \quad z_{0\ comp} = 0.$$

This means that the above composite micrometeorite (beryllium tube segment plus a graphite cone) will reach the Earth surface without being destroyed.

Conclusion

Artificial micrometeorites can be a convenient tool to study the Earth atmosphere.

Literature

1. Tables of physical quantities, Handbook ed. I. K. Kikoin, Moscow, Atomizdat, 1976

2. S.N. Dolya, KA Reshetnikova, On electrodynamics acceleration of macroscopic particles, Preprint, P9-2009-110, Dubna, 2009
http://www1.jinr.ru/Preprints/2009/110(P9-2009-110).pdf
http://arxiv.org/ftp/arxiv/papers/0908/0908.0795.pdf

3. S. N. Dolya, K. A. Reshetnikova, Linac ions C $^{+6}$ - injector synchrotron designed for hadrons therapy, Preprint JINR, P9-2011-82,
http://www1.jinr.ru/Preprints/2011/082(P9-2011-82).pdf
http://arxiv.org/ftp/arxiv/papers/1307/1307.6302.pdf

4. Physical quantities, Directory ed. I. S. Grigor'ev
and E. Z. Mey'likhov, Moscow, Energoatomizdat, 1991

5. I. M. Kapchinsky, Particle dynamics in linear resonance accelerators, Moscow, Atomizdat, 1966

6. S. N. Dolya, RF patents 2455800 and 2456782,
http://arxiv.org/ftp/arxiv/papers/1311/1311.5315.pdf

Appendix

Gas-dynamic acceleration of bodies till the hyper sonic velocity

The article considers an opportunity of gas-dynamic acceleration of body from the initial zero velocity till the finite velocity: V_{fin} = 5 km / s. When the gas flow rate of the body pre-acceleration reaches V_{in} = 1 km / s, the body is accelerated at the front of the explosion wave propagating along the coils of the hexogen spiral. This wave accelerates the body and, finally, it reaches the velocity of 5km/s. The accelerated body has mass m = 0.1 kg and diameter d_{sh} = 11.3 mm. Acceleration length is L_{acc} = 6 m. At the slope of the spiral to the horizon equal to Θ = 70^0 the flight range of the body is equal to: S_{max} = 1600 km, and the maximum height of the flight is H_{max} = 1100 km.

Introduction

There is a known [1] method of gas-dynamic acceleration of the body in the trunk. The powder disposed in the trunk for a short period of time transfers from the solid to the gaseous state, after that, due to its expansion it pushes the body out from the trunk. The finite velocity of the body in this method of the gas-dynamic acceleration is determined by the density and temperature of the composed gas.

Increasing the density of the gas and its temperature is limited by the durability of the trunk and can not be increased. Modern systems are designed for pressure 3500 atm. and temperature 3000 C^0. Further increase of the temperature and pressure beyond these values is not justified because at high temperatures the trunk is damaged and increase of the pressure requires the quality of the trunk to be improved. Further, at high temperatures there is an opportunity of dissociation of molecules of the gas that is related with the loss of the thermal energy of the gas. Therefore, at the present state of the technique development the achieved velocity of the body is the upper limit.

Taking into account the heat dissipation losses due to friction against the walls of the gas trunk and other losses it is possible to accept the limited velocity approximately equal to $V \approx 2$ km / s. This velocity is achieved when the body mass is a negligibly small part of the mass of the gunpowder. If you increase the mass of the body, the body velocity will seek to reach the ordinary velocity of the body: $V_{ord} \approx 1$ km / s.

To obtain the hypersonic velocity of the body about $V_{sh} \approx 5$ km / s is impossible by using this method.

In principle, one can accelerate the body by the explosion wave starting with the zero initial velocity of the body. At certain synchronization of the body with the explosion wave it is possible to accelerate the body at non- zero initial velocity. However, this method has a fundamental drawback: it is the small length interaction of the body with the explosion wave.

Let us consider an opportunity of accelerating the body till the hyper sonic velocity formed in the space by the explosion wave propagating synchronically with the accelerated body.

Selection of basic parameters

As the explosive substance we have chosen [2] the hexogen cord with a diameter $d_{cord} = 0.5$ cm, which is wound onto the tube with a diameter $2r_0 = 3$ cm. Let the body which is supposed to accelerate, be of a cylindrical shape with mass $m_b = 0.1$ kg, cross-section $S_{tr} = 1$ cm^2, and it has the initial velocity $V_{in} = 1$ km / s. The finite velocity of the body is chosen to be equal to $V_{fin} = 5$ km / s and the acceleration length is equal to $L_{acc} = 6$ m.

To obtain this acceleration, we find the following from the equation of motion of the body:

$$m_b dV / dt = F_{axis}, \qquad (1)$$

that the accelerating force must be equal $F_{axis} = 2 * 10^5$ N $= 2 * 10^5$ kg $*$ m/s^2. Indeed, dividing this force by body mass $m_b = 0.1$ kg, we find that the body will move with the uniformed acceleration: $a = 2 * 10^6$ m/s^2.

The velocity of the body will increase in accordance with the law:

$$V = V_{in} + at, \qquad (2)$$

and from the initial velocity $V_{in} = 1$ km / s till the finite velocity $V_{fin} = 5$ km / s the body will be accelerated during $t_{acc} = 2$ ms.

To implement the impact the permanently acting force onto the body, it is necessary to make an explosion of the hexogen cord wound on the carcass as a double spiral structure. The winding step must permanently grow so that the wave propagation velocity along the spiral would permanently coincide with the velocity of the body.

The dependence of the passed distance (at the uniformed accelerated motion) on the time can be written as follows:

$$L = V_{in}t + at^2 / 2 . \qquad (3)$$

Solving this equation for the velocity, we find that the dependence of the velocity on the covered distance is expressed as follows :

$$V (L) = (V_{in}^2 + 2 \, aL)^{1/2} . \qquad (4)$$

At the initial parameters, when $L = 0$, the velocity of the body is equal to $V = V_{in} = 1$ km / s. At the finite parameters: $L = L_{acc} = 6$ m, the velocity of the body is equal to: $V = V_{fin} = 5$ km / s.

Knowing the law of the velocity growth depending on the length of the passed distance, it is possible to find the dependence of the winding step on the passed distance.

Assume the velocity of propagation of detonation in hexogen to be equal to $V_{det} \approx 8$ km / s [2]. To have the velocity of the explosion wave V_{axis}, propagating along the axis of the spiral, permanently coincide with the velocity of the body V_b, the following relation must be satisfied:

$$V_b = V_{axis} \approx V_{det}h/2\pi r_0, \qquad (5)$$

where V_b - velocity of the body, V_{axis} - the velocity of the explosion wave propagating along the axis of the spiral, r_0 - radius of the spiral winding, h - the winding step of the spiral.

Detonation propagating along hexogen cord runs around the perimeter of the spiral, along the axis it runs with velocity $V_{axis} \approx V_{det}h/2\pi r_0$. Hence, for the initial velocity of the body, $V_{in} = 1$ km / s, the winding step should be equal to: $h_{in} \approx 2\pi r_0 V_{in}/V_{det} = 1.17$ cm, the finite step of winding of the spiral must be equal to: $h_{fin} \approx 2\pi r_0 V_{fin}/V_{det} = 5.88$ cm. Intermediate values of the winding step of the spiral are given by the following ratio:

$$h\,(L) = V\,(L) * V_{det}/2\pi r_0. \tag{6}$$

We will consider the detonation of the double spiral with a diameter of each of the hexogen cords equal to d_{cord}, equal to: $d_{cord} = 0.5$ cm.

Pressure on the detonation front in hexogen reaches $P_{hex} \approx 30$ GPa [2]. Let the body be located at the distance equal to the radius of the spiral winding $z = z_s = 1.5$ cm at the moment when the explosion wave comes up to its back slice.

We will consider the detonation of the long cylindrical hexogen cord. In the spread out of the explosion products the pressure of the explosion wave P_{wave} at the back slice of the body would have been less than the pressure on the front of the wave by the ratio of the square of the radius of the hexogen cord: $(d_{cord} / 2)^2 = 0.625$ cm² to the square of the distance till the back slice of the body (4.5 cm²). It would have been as follows: $P_{wave1} = P_{hex}r^2_{cord} / (r_0^2 + z_s^2) = 0.4$ GPa. Since the construction of detonating chamber is like this, the explosion products leave the area of the explosion only through one quadrant, the pressure at the front of the explosion wave is by 4 times more than in the free space and equal to $P_{wave2} = 1.6$ GPa.

In this case the force acting on the back slice of the body is as follows:

$$F_{axis} = P_{vawe2} * S_{tr} * \cos\varphi, \tag{7}$$

where we have chosen : $\varphi = 45^0$, $\cos\varphi = 0.7$. Taking into account that two cords are simultaneously detonated, their forces are summed up.

Substituting the numerical values into formula (7) we find that this force is equal to $F_{axis} \approx 2.2 * 10^5$ N.

The light carcass with the hexogen spiral and rings of the special shape, as well as the separable cylinder are located inside the strong trunk with the internal diameter of D_{bar} = 80 mm. In this case the pressure onto the inner surface of the trunk at the absence of the substance between the strong trunk and hexogen cord would have been equal to the value:

$P_{bar} = P_0 * (d_{cord} / 2)^2 / [(D_{bar} / 2) - r_0]^2 \approx 3000$ atm., which is the ordinary pressure for the guns [1]. Due to the presence of the rings and separable cylinder the pressure on the inner surface of the strong trunk will be less.

After reaching the velocity V = 2 km / s it is possible to use a quadrifilar spiral. But it is necessary to reduce the diameter of the two cords accordingly (by the root of 2 times). When velocity V = 3 km / s it will be possible to use the six-filar spiral, and etc.

Synchronization of the body being accelerated and the explosion wave

As it is shown in [3], the velocity of detonation product spread out of the cylindrical cord in the transverse direction is: $V^{\perp} = 0.8 V_{det}$, i.e. In our case this velocity is approximately equal to $V^{\perp} \approx 6.5$ km / s. Detonation products should reach the back slice of the body at the moment when the back slice is at the distance equal to the radius of the spiral - $l_{in} = r_0 = 1.5$ cm. Then the distance from the back slice of the body to the explosive area of the detonating cord must be $l_{body} \approx 2.12$ cm.

Explosion products fly over this distance during the time τ_{shok} equal to: $\tau_{shok} = l_{body} / V^{\perp} = 3.26$ µs. The body moving with the initial velocity $V_{in} = 1$ km / s, must be at this moment at the distance $l_{in} = 3.26$ mm from the start of the acceleration.

Phase stability principle for the body being accelerated on the front of the explosion wave

As in particle accelerators on the traveling wave there is the phase stability [4] while accelerating the body at the front of the explosion wave. In the particle accelerators the acceleration phase for particles is chosen in advance, it is called synchronous. For this phase the longitudinal motion of the particles is calculated after that. If any particle is accidentally behind the synchronous phase, it will get into the stronger electric field. Then it will be more accelerated and finally will catch up with the synchronous phase.

If the particle is too fast in its motion for the synchronous phase, then it will get into a smaller electric field, it will be less accelerated, and, finally, the accelerated moving pulse and the synchronous phase will catch up with the particle.

Below we show that in this case while accelerating the body at the front of the explosion wave, the dependence of the force acting for the body on the phase of the pulse at its front, has a declining character.

The force acting on the body, $F_{axis} = P_{wave} * S_{tr} * \cos \varphi$, depends on the pressure $P_{wave} = P_{hex} * r_{cord}^2 / (r_0^2 + z^2)$, where $P_{hex} = 30$ GPa - pressure of the explosion wave (hexogen), $2r_{cord} = 0.5$ cm - diameter of the hexogen cord, $r_0 = 1.5$ cm – the hexogen cord winding radius, z - the distance along the axis from detonating area till the back slice of the body (cm). Projection of the pressure force on the axis of the acceleration is proportional to $\cos \varphi$, which can be represented as follows: $\cos \varphi = z / (r_0^2 + z^2)^{1/2}$.

Thus, the dependence of the force acting on the body along the axis (F_{axis}) on the distance along the axis between the detonating area and the back slice of the body, is expressed as follows:

$$F_{axis} \sim z / (r_0^2 + z^2)^{3/2}. \qquad (8)$$

Below we compile a table of the values of this function for three values of the distance between the detonation region and the back slice of the body.

Table 1. Values of the function $z / (r_0^2 + z^2)^{3/2}$

z, cm	$z/(r_0^2+z^2)^{3/2}$
1	0.17
1.5	0.157
2	0.128

From the analysis of this function it is clear that the force acting on the body reduces if the body is faster than the synchronous phase. The explosion force acting on the body increases if the body is behind the synchronous phase. This dependence corresponds to the stable phase of the longitudinal motion according to the phase stability principle.

Transverse movement of the body

Let us consider the transverse force available at the front of the explosion wave on the body. In the beginning of the acceleration the velocity of the explosion wave propagating along the axis of the spiral is $V_{in} = 1$ km / s. The detonation velocity propagating along the spiral is $V_{det} = 8$ km / s. The perimeter of one of the spiral coils is $\pi d_{spir} \approx 10$ cm, so that the time of detonation of the total cycle is as follows: $\tau_{cyc} \approx \pi d_{spir} / V_{det} = 10$ µs.

Assume that the body moves transversely accelerated but by two orders of the magnitude lower than in the longitudinal direction, i.e., $a = 2 * 10^4$ m/s². During the time when the detonation runs one third of the total cycle $\tau_{cyc1/3} = 3$ µs, the body is shifted by a distance of $S_{1/3} = a \tau^2_{cyc\ 1/3}/2 = 0.1$ µ. After that the transverse force changes its direction and after the total cycle its action is averaged.

Ballistics. Aerodynamic resistance

Now we calculate the motion of the body released at an angle of $\Theta = 70^0$ to the horizon, taking into account the air resistance. The equation of the horizontal motion of the body can be written as follows:

$$m_b\, dV_x / dt = \rho C_x S_{tr}\, V_x^2 / 2, \qquad (9)$$

where m - mass of the body , V_x- horizon velocity, g - 0.01 km/s² - acceleration due to gravity , $\rho = \rho_0 e^{-z/H0}$ - barometric formula of the change of the atmospheric density in dependence on the height , $\rho_0 = 1.3 * 10^{-3}$ g/cm³ - air density at the Earth surface, $H_0 = 7$ km - the height at which this density decreases by a factor of e.

The aerodynamic coefficient or coefficient of drag resistance is presented as a dimensionless quantity, taking into account the "quality" of the body shape:

$$C_x = F_x / (½)\, \rho_0 V_x^2 S_{tr}. \qquad (10)$$

The solution of equation (9) can be written as follows:

$$V(t) = V_x / [\, 1 + \rho C_x V_x * S_{tr} * t/2m_b\,]. \qquad (11)$$

In order to calculate the change in velocity of the body in dependence on time, you need to find the aerodynamic coefficient C_x.

Calculation of the aerodynamic coefficient for the air

We assume that the body has the shape of a cylindrical rod with a conical head. Then at the hit of a nitrogen molecule on the sharp cone, the change of the longitudinal velocity of the molecules is equal to:

$$\Delta V_x = V_x * \Theta_t^2 / 2, \tag{12}$$

where Θ_t –the cone angle at the vertex. Gas molecules transfer the momentum to the body:

$$p = mV = \rho V_x S_{tr} t * \Delta V_x. \tag{13}$$

The change in the momentum per unit of the time is the force which is called the force of the frontal slow down,

$$F_{x1} = (\tfrac{1}{2}) \rho V_x S_{tr} * V_x * \Theta_t^2. \tag{14}$$

Dividing F_{x1} by $(\tfrac{1}{2}) \rho V_x^2 S_{tr}$, we obtain the drag coefficient for the sharp cone at the mirror reflection of the molecules from the cone, (Newton's formula):

$$C_{x\ air} = \Theta_t^2. \tag{15}$$

Our consideration corresponds to hypersonic velocity; we can neglect the effects that occur at the velocity close to the sonic velocity in the unperturbed medium. Let the length of the conical part of the body be as follows: $l_{cone} = 65$ mm and the diameter of the body $d_b = 11.3$ mm. This means that the angle at the vertex of the cone is: $\Theta_t = 0.173$ and $C_{x\ air} = 0.03$.

Substituting this value in the formula for the C_x loss of the longitudinal velocity in dependence on time (11), we find that the decrease in the vertical velocity of the body in the first second of the flight is of the order of 10%.

The flight range of the body is $S_{max} = 2V_0^2 * \sin\Theta * \cos\Theta / g = 1600$ km, the maximum lifting height of the body is $Y = V_0^2 * \sin^2\Theta/2g = 1100$ km.

Fig. 1 shows a diagram of the device.

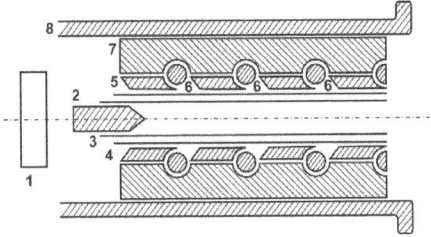

Fig.1. 1 – the gun, 2 - the body, 3 – the light trunk, 4 - light carcass, 5 - special shaped rings, 6 - hexogen cord, 7 - separable cylinder, 8 - solid trunk.

The operation of the device is as follows. In the gun (1), the acceleration of the body (2) of a cylindrical shape with mass $m_b = 0.1$ kg, and the cross-section of $S_{tr} = 1$ cm^2 is performed by using the ordinary gas - dynamic method. The body is centered relatively the spiral with a light trunk (3). The body moves in the light carcass (4), where the special shaped rings (5) are embedded. Synchronously with the start of the acceleration we produce the explosion of the hexogen cord (6), which is placed in the gaps between the rings. Outside the cord is surrounded with the separable cylinder (7). The rings must be rigidly fixed with the separable cylinder. Figure 1 does not show fastening of the rings to the cylinder. All the assembly is placed inside the solid trunk (8). Detonation propagates along the cord and creates the explosion wave running synchronously with the body.

If to increase the mass of the body by an order of the magnitude till the value of $m_{b1} = 1$ kg, then the length of the acceleration (at the same parameters) will increase by the order of magnitude and will be equal to the value of $L_{accl} = 60$ m. Such an accelerator may be arranged only horizontally, while at the same time the body must have asymmetry which creates the lifting force.

Suppose that the lifting coefficient is, $C_y = 0.2$. In this case the drag coefficient must be the same: $C_x = 0.2$.

Thus, the equation of the vertical motion can be written as follows:

$$m_{b1}dV_y / dt = C_y \rho_0 V_x^2 * S_{tr} / 2, \qquad (16)$$

where C_y is the aerodynamic lifting force coefficient, $\rho_0 = 1.3 * 10^{-3}$ g/cm^3 - the air density on the surface of the Earth, $V_x = 5$ km / s - the horizontal velocity of the body, S_{tr} is the cross-section of the body.

Solving approximately equation (16), we obtain:

$$V_y = C_y \rho V_x^2 S_{tr} t / 2m_{b1}. \qquad (17)$$

Integrating again, we obtain an expression for the lifting height of the body:

$$Y_1 = C_y \rho V_x^2 S_{tr} t^2 / 4m_{b1}. \qquad (18)$$

Solving the equations of horizontal and vertical movements we can obtain the time dependence of the flight parameters of the body, which we present in Table 2. The first column gives the time of the flight, the second - the horizontal velocity, in the third – there is the vertical velocity, the fourth column shows the height of the lifting of the body.

Table 2. Flight parameters for the case of C_x, $C_y = 0.2$.

t, s	V_x, km/s	V_y, km/s	Y, km
0	5	0	0
2	4.43	0.65	0.65
5	3.8	1.37	2.7
10	3	2.3	8.5
20	2.68	2.66	20

The time of the body lifting till the maximum height in this case is as follows: $\tau_{max} = V_y / g = 266$ s. The flight range of the body is $S_{max1} = V_x * 2\tau_{max} = 1400$ km, the maximum body lifting height is $Y_1 = V_y^2 / 2g = 350$ km. Changing the shape of the cone in the head part of the body, it will be possible to obtain different trajectories.

Conclusion

These flight parameters may be of interest for a number of applications, in particular, for the removal of the space garbage at the low orbits.

Literature

1. I. A. Sterznev, Artillery guns multiple actions, the speed limit of artillery shells, http://ivanstrezhnev.appspot.com/3/3.html

2. http://ru.wikipedia.org/wiki/Гексоген

3. V.V. I'lyin, A.P. Rybakov, V.V. Kozlov, Mathematical model of expansion of explosion products at outlet oblique detonation wave at the free surface, Electronic scientific journal "INVESTIGATED IN RUSSIA", 2006, p. 1531, http://zhurnal.ape.relarn.ru/articles/2006/165.pdf

4. Veksler, Reports of the USSR, v. 43, issue 8, p. 346, 1944
   E. M. McMillan, Phys. Rev., V. 68, p. 143, 1945

www.ingramcontent.com/pod-product-compliance
Lightning Source LLC
Chambersburg PA
CBHW081129170526
45165CB00008B/2610

9 781500 842017